Solar and Renewable Energy -The future Resources of Power and Electricity

By

Sahriar Ahamed

Table of Contents

About the Author:..4

Dedication:...6

Fuel generating electricty systems & awareness:............8

Making popular to use alternative energy by WB:..........15

Manufacturing process of PV cell:....................................16

Characteristics of PV Module:...25

How the solar cell works:...30

Diffrent types of solar electric systems and How they can be built
    up:...39

Grid-connected PV systems...48

Maximum Power Point Tracker (MPPT):...........................51

Charge Controllers in PV Systems:...................................53

Storage System:...58

About the Author:

The author is a global freelance article writer and has done commerce graduation twice. He completed Masters Degree twice; one in management and the other in marketing. He has achieved professional training and diploma certification on Solar and Renewable energy equipments sector from top most University in Bangladesh, University of Dhaka. By profession he is working as business development manager in a reputed Solar and Renewable energy providing company in Bangladesh and also helping indirectly World Bank to set up off grid electricity connection in remote area for the deprived people of Bangladesh. He is married and a father of a wonderful daughter the whole family leading a very simple life in Dhaka Bangladesh

Sahriar Ahamed

Dedication:

This book is dedicated to

1. You Readers who by purchasing this book started to change the world from relying on Fossil fuels to Renewable energy

2. To the sincere efforts made by Renewable Energy Enthusiasts,

    Solar Energy Providers

3. UN and Governments who are involved in the transformation of the world to renewable energy

Very special dedication and thanks to

Author's family

Hridita & Roshan Ara

Who is this book meant for:

In reviewers' opinion as this is written by an author who is practically involved in installing alternate energy systems for many years

1) General Readers as they are using power in different form everyday

2) Students at any level of Science and Engineering especially Physics, Environment, Electrical, Automotive

3) Governments and Electricity Power providers

4) Automobile and Aircraft manufacturers

What will happen when the fossil fuel and other resources of power and electricity end within few years?

With the current world situation, this question has the most logical base. With the reserves of crude oil just over 1000 billion barrels that is available for usage, the world can run only for another 32 (Thirty Two) years. Where will our children draw energy from? When we understand that the above calculation is based on the current utilization of 84.1 million barrels a day, and at the rate with which we increase our consumption everyday it won't be out of context to modify our calculation of consumption to double in few years then the stock will last for just 16 years, now forget about our children and the fuel will not last even for our consumption in our later years. What will happen to Cars, Boats, and Flights? Will all of them be abandoned? Yes as well as No are two alternate answers come up based on what we do from here on.

After the stock report of fossil fuel; let's check out where this fuel is being utilized: -

- The main consumption of fossil fuel is to produce power and electricity.

- The second most usage is to run vehicle and to run a smooth transportation system too.

- Other than these: mills, factories, heavy industrial instruments and equipments are run by the fuel.

So, it is natural to come to our mind what will happen after 16 or 32 years from now when we will finish off all the reserve of oil. What a pity at that time we fully would have exhausted a bounty given to us by Nature to improve our lives, Then the afore mentioned questions put up another time for emphasise

➢ Will we stop our power and electricity production?
➢ Will we stop the heavy industries?
➢ Will we stop the transportation and vehicle?

Anyone can guess that is not possible for us. World can't pass a second without power and energy in the modern times. So, there is usage of alternative and renewable energy resources which we are planning to make as the substitution to the fossil fuel and the coal & natural gas. The watchword, the sustainable renewable energy is getting popular as clean and green energy. Most of the countries are investing on this natural and renewable energy sector. There is no other way left for the human civilization ahead, other than this. Global and Local Energy Scenarios are presented that the scenarios give important messages to decision makers in political, technological, economic and industrial sectors in the energy field. The energy scenarios have a critical role to develop a sustainable energy framework and assessment of the factors and issues. Three Scenarios up to 2050 have been chalked out; A reference scenario, energy revolution scenario and an advanced energy revolution version. The reference Scenario is based on WEO assuming average annual growth rate GDP of 3.1% from 2007-2030, energy consumption in 2030, 6%, Global Purchase power parity 29% by 2050. The energy revolution is based on the reduction of CO_2 around 10 Gton by 2050. The advance revolution theory assumes a much less CO_2 considering the use of renewable energy resources and technologies like SPV, wind, CSP etc. The aim of the scenario is to make a change in the way the energy is produced, distributed and consumed. The principles behind this change will be. Natural limits of the environment must be respected. Renewable solutions must be implemented through decentralized energy systems. The aim of the scenario is to make a change in the way the energy is produced, distributed and consumed.

The principles behind this change will be

1) Natural limits of the environment must be respected.
2) Renewable solutions must be implemented through decentralized energy systems.
3) The use of resources should be created.
4) Phase out unsustainable energy sources.
5) Decouple economic growth from the consumption of fossil fuels.

Energy can be classified into several types based on different criteria such as Primary and Secondary energy, Commercial and Non commercial energy and Renewable and Non-Renewable energy. Primary energy sources are those that are either found or stored in nature. Common primary energy sources are coal, oil, natural gas, and biomass (such as wood). Other primary energy sources available include nuclear energy from radioactive substances, thermal energy stored in earth's interior, and potential energy due to earth's gravity. The major primary energy sources are mostly converted in industrial utilities into *secondary energy* sources; for example coal, oil or gas converted into steam and electricity. The energy sources that are available in the market for a definite price are known as commercial energy. By far the most important forms of commercial energy are electricity, coal and refined petroleum products. Commercial energy forms the basis of industrial, agricultural, transport and commercial development in the modern world. In the industrialized countries, commercialized fuels are predominant source not only for economic production, but also for many household tasks of general population. Examples: Electricity, lignite, coal, oil, natural gas etc. The energy sources that are not available in the commercial market for a price are classified as non-commercial energy. Non-commercial energy sources include fuels such as firewood, cattle dung and agricultural wastes, which are traditionally gathered, and not bought at a price used especially in rural households. These are also called traditional fuels. Non-commercial energy is often ignored in energy accounting. Example: Firewood, agro waste in rural areas; solar energy for water heating, electricity generation, for drying grain, fish and fruits; animal power for transport, threshing, lifting water for irrigation, crushing sugarcane; wind energy for lifting water and electricity generation. Renewable energy is energy obtained from sources that are essentially inexhaustible. Examples of renewable resources include wind power, solar power, geothermal energy, tidal power and hydroelectric power. The most important feature of renewable energy is that it can be harnessed without the release of harmful pollutants. Non-renewable energy is the conventional fossil fuels such as coal, oil and gas, which are likely to deplete with time. POLICY target is fixed up as: Peak global temperature rise well below 2°C. Reduce GHG emissions by 40% by 2020 (as compared to 1990) in developed countries. Reduce GHG emissions by 15 to 30% of projected growth by 2020 in developing countries. Achieve zero deforestation globally by 2020. Agree a legally binding global climate deal as soon as possible. The use of resources should be created. Phase out unsustainable energy sources. Decouple economic growth from the consumption of fossil fuels. The next generation survival is depending on the perfection of proper utilization of invested capital in research of sustainable and renewable energy resources production and implementation system. USA president Mr. Barak Obama called for the country to reduce its oil import by one third by the year 2025. This is not that they are in short of budget for buying oil from other countries but they would like to secure its energy supply in the future. May be question will arise how? We can mention Mr. Obama's speeches here,"When I was elected to this office, America imported 11(eleven) million barrels of oil a day. By little more than a decade from now we will cut that by one third". Actually his emphasis on using of alternative of oil and fossil fuel; he said "The only way for American energy supply to be truly secure is by permanently reducing our dependence on oil. We are to have to find ways to boost our efficiency so we use less oil. We have got to discover and produce cleaner renewable sources of energy that also produce less carbon pollution". He said that meeting this new goal of cutting US oil dependence depends largely on two things finding and producing more oil at home and reducing our dependence on oil with cleaner alternative fuels and greater efficiency. The USA president pointed on not to make a quick fix of the solution. Here we can mention another project that is completely hydro and the location is in Himachal. This project is established by a consortium of TATA Power and Norwegian hydro power company. This project is build to produce 236 Mega Watt of electricity in Chenab Valle in Himachal Pradesh, India. In Sri Lanka, rural electricity schemes has launched very recently. Several projects are going to be established by the government of Sri Lanka to achieve the target of 100 percent electricity needs of country's rural sectors by 2012 set by Minister Patali Chapika Ranwake. The 4th is a lighting project in Sri Lanka, Ranpura, Kague, Central province. The rural electricity project No.8 issued by the power and energy commission is the project under which new electricity connection had been provided for 720 houses in the Galle and Matara districts. Another 516 electricity schemes of Rs. 85.7 million are in operation covering Elipitya, Nagoda, Karon, Deniya, Yakkalamulla,Beddegama and Thavalama divisions in the Gale district, while 203 electricity schemes will be started at the cost of Rs. 134.1 million to cover Kotapola, Pitabed, Passara, Dickweili and Akuresra DS divisions in Matara district. The International Energy Agency 794(IEA) is saying that the wide spread development of "Smart Grids" networks that monitors and manage the transport of electricity from all sources to meet the varying electricity demands of end users is crucial to achieving a more secure and sustainable future. The worldwide Photo voltaic (PV) market has been growing at over 35% per annum in recent years and it can now make a significant contribution to electricity generation Solar PV is a critical part of the energy mix - it can be used in decentralized or centralized formats, it is useful in an urban environment and has huge potential for cost reduction. Concentrating solar power is currently experiencing massive expansion, and costs are expected to be 6 to 10 ct/kWh in the long term. Solar thermal 'concentrating' power stations (CSP) are suitable for areas with high levels of direct sunlight. The technical potential of North Africa for CSP, for example, is much greater than local demand. There is more than enough solar radiation available all over the world to satisfy a vastly increased demand for solar power systems. The sunlight, which reaches the earth's surface, is enough to provide 2,850 times as much energy as we can currently use. On a global average, each square meter of land, if exposed to enough sunlight, can produce 1,700 kWh of power every year. The average irradiation in Europe is about 1,000 kWh per square meter, however, compared with 1,800 kWh in the Middle East..

Making popular to use alternative energy by WB:

Due to the fact the World Bank has taken an initiative to allocate the loan and subsidies in the sector of renewable energy nowadays, the IMF has prepared fund for set up the renewable energy plant, such as; the electricity production by the photo voltaic (PV) solar cell. Wind mill and air generated electricity; hydro electricity; bio-gas plant and so on. From a source it is known that World Bank will allocate a fund for Bangladesh, a country of East Asian continent. For first 20,000 solar home systems (SHS) World Bank will grant $90 for each solution system; for next 20,000 system $ 70 and for the next 30,000 system is $50 each. Not only that; GTZ granted a fund for the system of 33,660 systems. They are paying €38 for each solar home system (SHS). KfW funds had grant €38 each for first 30,000 systems and €36 each for next 35,000 systems and €34 each for Next 35,000 systems. So, anyone can get the idea how profitable it is to install the solar home system in Bangladesh. In a fact analysis it has been seen that suddenly the business of solar home system (SHS) has been groomed and increased at a faster rate. Within a few years there are several companies and individuals who are investing money in this sector not only to have the fund allocated by the international organization but also to capture the current market share also. This is just one example in a third world country in the continent of Asia. There are several countries like Bangladesh in the continent of Asia which have poor and underdeveloped economies. Other than Asia, Africa and Latin America also face the same problem. These under developed countries are getting fund from international organizations and the developed countries of Europe and USA are arranging fund of themselves. Now the reader may get the picture very clear of renewable and alternative energy market worldwide.

Now we can discuss about the of a SOLAR PHOTOVOLTAIC (PV) technology. The manufacturing process of solar photo voltaic (PV) panels can be broken down into 10 (ten) steps. They are illustrated in the following image:

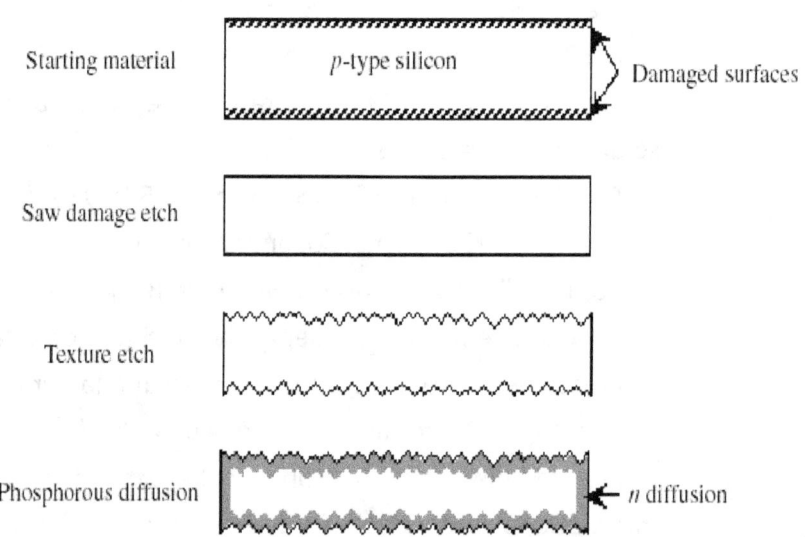

A typical processing sequence with schematic illustrations of the resulting structures

1. ***Starting material:*** The industry uses so-called solar grade Cz-Si wafers, round in origin but very often trimmed to a pseudo-square shape, or multicrystalline square wafers. Wafer dimensions are between 10- and 15-cm side and between 200- and 350-µm thick. Doping with *p*-type (boron) resistively to 1 $\Omega \cdot$cm.

2. ***Saw damage removal***: The sawing operation leaves the surfaces of "as cut" wafers with a high degree of damage. This presents two problems: (a) the surface region is of a very bad quality and (b) the defects can lead to wafer fracture during processing. For this reason, about ten microns are etched off from each face in alkaline or acid solutions. The wafers, in Teflon cassettes, are immersed in tanks containing the solution under temperature and composition control. Alkaline etches are preferred to acid solutions due to considerations of waste disposal.

3. ***Texturization***: NaOH etching leading to microscopic pyramids is commonly employed. Their size must be optimized, since very small pyramids lead to high reflection, while very large ones can hinder the formation of the contacts. To ensure complete texturing coverage and adequate pyramid size, the concentration, the temperature, the agitation of the solution and the duration of the bath must be controlled (in fact NaOH at a higher concentration and at a higher temperature is commonly used as an isotropic etch for saw damage removal). Alcohol is added to improve homogeneity through an enhancement of the wettability of the silicon surface. Typical parameters are 5% NaOH concentration, 80°C and 15 min.

4. ***Phosphorus diffusion***: Phosphorus is universally used as the *n*-type dopant for silicon in solar cells. Since solid-state diffusion demands high temperature, it is very important that the surfaces are contamination-free before processing. To this end, after the texturing the wafers are subjected to acid etch to neutralize alkaline remains and eliminate adsorbed metallic impurities. *Quartz furnaces*: The cells to be diffused, loaded in quartz boats, are placed in a quartz tube with resistance heating and held at the processing temperature

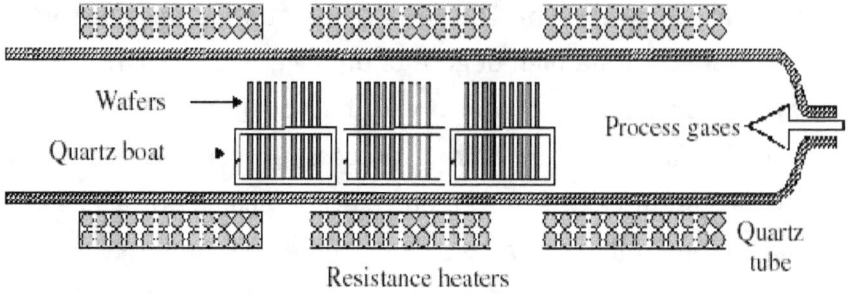

Fig: Quartz furnaces

The cells enter and exit the furnace through one end, while gases are fed through the opposite one. Phosphorus itself can be supplied in this way, typically by bubbling nitrogen through liquid POCl3 before injection into the furnace. Solid dopant sources are also compatible with furnace processing. Five to fifteen minutes at temperatures in the range from 900 to 950.C can be considered representative; both surfaces and the edges of the wafer will be diffused.

5. ***Junction isolation***: The *n*-type region at the wafer edges would interconnect the front and back contacts: the junction would be shunted by this path translating into a very low shunt resistance. To remove this region, dry etching, low temperature procedures are used. For the widely employed etch with plasma, the cells are coin-stacked and placed into barrel-type reactors. In this way, the surfaces are protected and only the edges remain exposed to the plasma. This is obtained by exciting with an RF field a fluorine compound (CF4, SF6), which produces highly reactive species, ions and electrons that quickly etch the exposed silicon surface. Laser cutting of the wafer edges is an alternative   in   industrial   use.

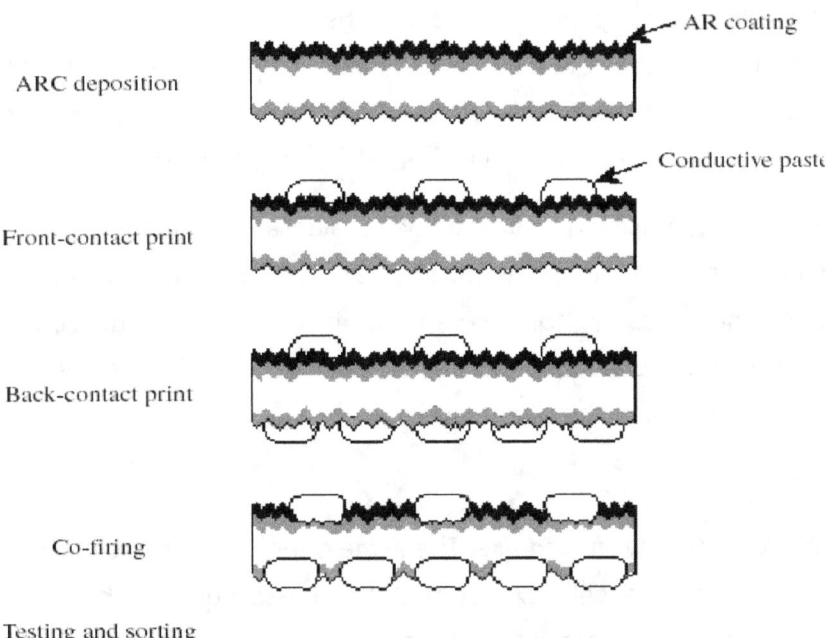

ARC deposition

Front-contact print

Back-contact print

Co-firing

Testing and sorting

Fig: A typical processing sequence with schematic illustrations of the resulting structures

6. **ARC deposition**: Titanium dioxide (TiO2) is often used for creating the antireflection coating due to its near-optimum refraction index for encapsulated cells. A popular technique is atmospheric pressure chemical vapor deposition (APCVD) from titanium organic compounds and water: the mixture is sprayed from a nozzle on the wafer held at around 200°C and the compound is hydrolyzed on the surface. This process is easily automated in a conveyor-belt reactor. Other possibilities include spin-on or screen print appropriate pastes. Silicon nitride is also used as AR coating material with unique properties.

7. **Front contact print and dry**: The requirements for the front metallization are low contact resistance to silicon, low bulk resistively, low line width with high aspect ratio, good mechanical adhesion, solderability and compatibility with the encapsulating materials.

   Resistively, price and availability considerations make silver the ideal choice as the contact metal. Copper offers similar advantages, but it does not qualify for screen printing because subsequent heat treatments are needed during which its high diffusivity will produce contamination of the silicon wafer.

8. **Back contact print and dry**: The same operations are performed on the backside of the cell, except that the paste contains both silver and aluminum and the printed pattern is different. Aluminum is required because silver does not form ohmic contacts to p-Si, but cannot be used alone because it cannot be soldered. The low eutectic temperature of the Al-Si system means that some silicon will be dissolved and then recrystallize upon cooling in a p-type layer.

9. ***Coffering of metal contacts***: A high-temperature step is still needed: organic components of the paste must be burnt off, the metallic grains must sinter together to form a good conductor and they must form an intimate electric contact to the underlying silicon. As Figure 7.6 shows, the front paste is deposited on an insulating layer (the AR coating) and the back contact on the parasitic n-type rear layer.

10. ***Testing and sorting***: The illuminated $I - V$ curve of finished cells is measured under an artificial light source with a spectral content similar to sunlight (a sun simulator). In order to obtain adequate output voltage, PV cells are connected in series to form a PV module. Since PV systems are commonly operated at multiples of 12 volts, the modules are typically designed for optimal operation in these systems. The design goal is to connect a sufficient number of cells in series to keep Vm of the module within a comfortable range of the battery/system voltage under conditions of average irradiance.

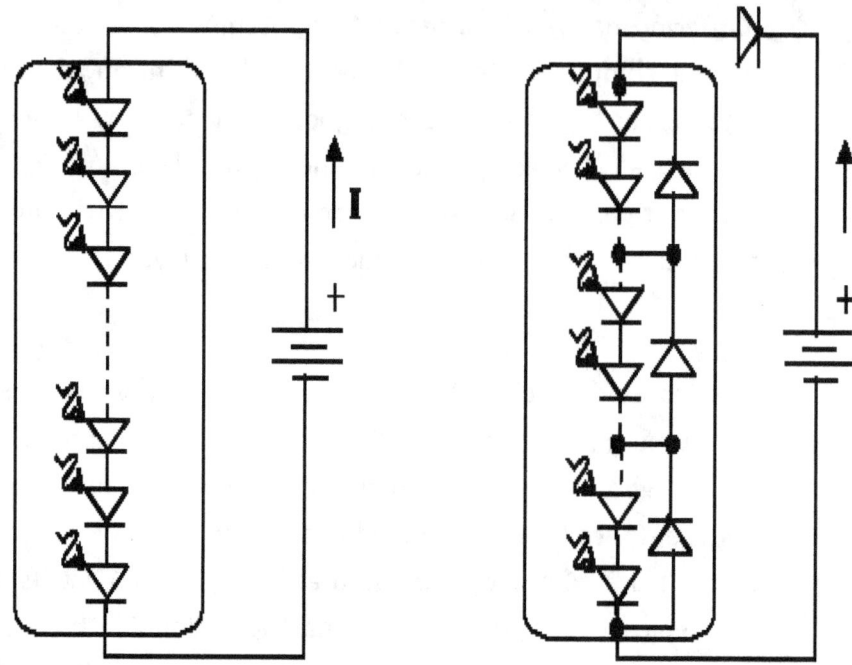

lule without blocking or bypass diodes   b. Module with blocking and bypass diode

gure 3.7 Battery discharge path through PV module with and without blocking diode.

If this is done, the power output of the module can be maintained close to maximum. This means that under full sun conditions, Vm should be approximately 16–18 V. Since V m is normally about 80% of VOC, this suggests designing the module to have a VOC of about 20 volts. With silicon single cell open-circuit voltages typically in the range of 0.5– 0.6 volts, this suggests that a module should consist of 33–3 6 cells connected in series. With each individual cell capable of generating approximately 2–3 watts, this means the module should be capable of generating 70-100 watts. Under night-time conditions, when none of the cells are generating appreciable photocurrent, it is necessary to consider the module as a series connection of diodes that may be forward biased by the system storage batteries. For example, suppose the module consisted of 33 cells, each of which has a reverse saturation current of 10–10 A. Suppose also that the system battery voltage is 12.8 volts, and that this voltage is uniformly distributed across the series cells. This means that each cell will have 0.388 volts across it in the forward bias direction. Diode equation shows this will result in a current of 0.32 mA flowing from the batteries into the PV module. If the module consisted of 28 cells, this might result in more efficient battery charging during peak sun, since Vm would be approximately 14.6 V, which is closer to the battery voltage. However, under no sun, each cell would have 0.457 V across it and a battery discharge current of 4.63 mA would flow. Furthermore, under weak sun or high temperatures, the module output voltage could be less than the battery voltage and no charging would occur. If another diode is connected in series with the module to prevent current from flowing in the reverse direction, this **blocking diode** will then have a forward voltage drop and associated power loss of more than 1 watt when the module is providing photocurrent. If the module is only providing 50 watts, this loss represents 2–3% of the total module output power. Another important observation relating to the series connection of PV cells relates to shading of individual cells. If any one of the cells in a module should be shaded, the performance of that cell will be degraded. Since the cells are in series, this means that the cell may become forward biased if other unshaded modules are connected in parallel, resulting in heating of the cell. This phenomenon can cause premature cell failure. To protect the system against such failure, modules are generally protected with **bypass diodes**, as shown in Figure 3.7. If PV current cannot flow through one or more the PV cells in the module, it will flow through the bypass diode instead. When the PV cells are mounted in the module, they can be characterized as having a **nominal operating cell temperature** (NOCT). The NOCT is the temperature the cells will reach when operated at open circuit in an ambient temperature of 20oC at AM 1.5 irradiance conditions, G = 0.8 kW/m2 and a wind speed less than 1 m/s. For variations in ambient temperature and irradiance the cell temperature (in oC) can be estimated quite accurately with the linear approximation. The combined effects of irradiance and ambient temperature on cell performance merit careful consideration. Since the open circuit voltage of a silicon cell decreases by 2.3 mV/$°C$, the open circuit voltage of a module will decrease by 2.3n mV/ $°C$, where n is the number of series cells in the module. Hence, for example, if a 36-cell module has a NOCT of 40oC with VOC = 19.40 V, when G = 0.8 kW/m2, then the cell temperature will rise to 55oC when the ambient temperature rises to 30oC and G increases to 1 kW/m2. This 15oC increase in cell temperature will result in a decrease of the open circuit voltage to 18.16 V, a 6% decrease.

Basically Photovoltaic (PV) or solar cells are PN junction Semiconductor devices. It converts sun light into direct current electricity. There are some types of solar cells or PV solar cells in market: such as, Mono Crystalline Solar cell: The essence of this process is the use of a semiconductor material which can be adapted to release electrons, the negatively charged particles that form the basis of electricity. The most common semiconductor material used in photovoltaic cells is silicon, an element most commonly found in sand. Crystalline silicon cells are made from thin slices cut from a single crystal of silicon is mono crystalline.

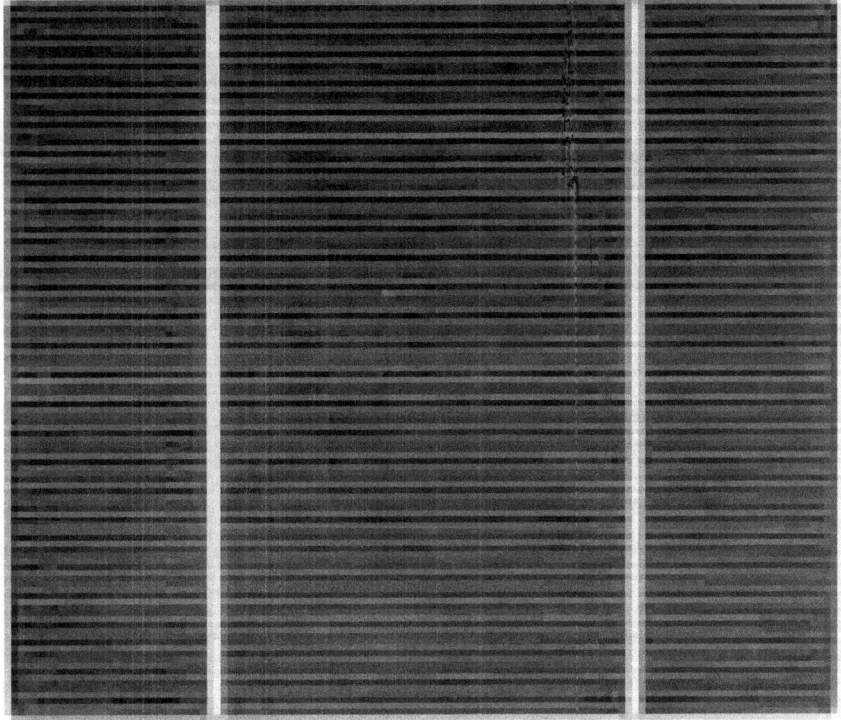

Picture: **Mono-crystalline solar cells**

Poly Crystalline Solar Cell: From a block of silicon crystals (polycrystalline or multi crystalline). This is the most common technology, representing about 80% of the market today; In addition, this technology also exists in the form of ribbon sheets.

Picture: Poly-crystalline solar cells

Thin film modules are constructed by depositing extremely thin layers of photosensitive materials onto a substrate such as glass, stainless steel or flexible plastic.

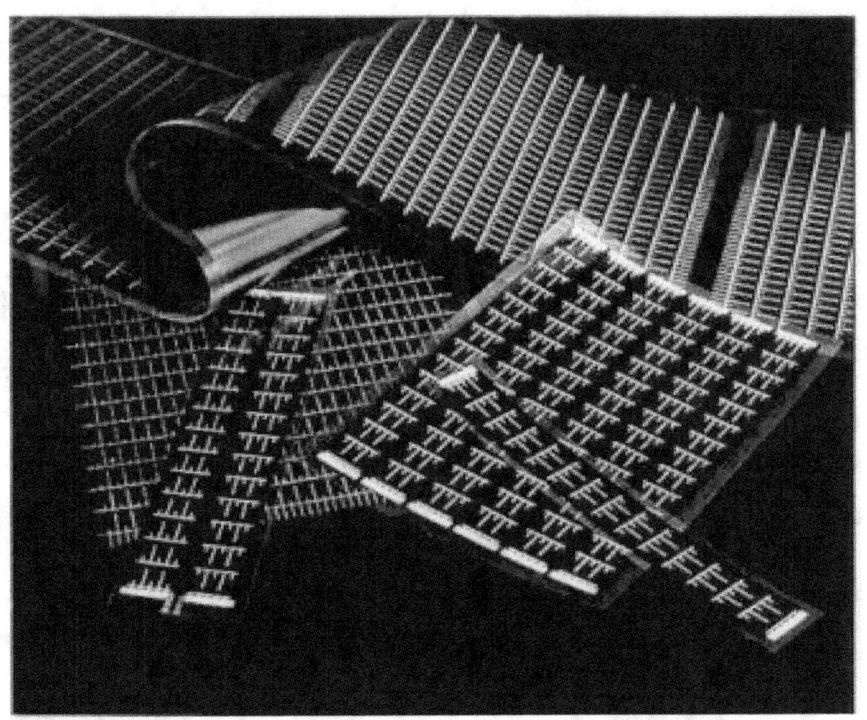

Picture: Flexible amorphous silicon

The latter opens up a range of applications, especially for building integration (roof tiles) and end-consumer purposes. Four types of thin film modules are commercially available at the moment: Amorphous Silicon, Cadmium Telluride, Copper Indium/Gallium Dieselenide/Disulphide and multi-junction cells.

Picture: Thin film photovoltaic panels being installed onto a roof

Picture: Sport car covered with plastic solar cells

Other emerging cell technologies (at the development or early commercial stage):

These include Concentrated Photovoltaic ,(CSP) consisting of cells built into concentrating collectors that use a lens to focus the concentrated sunlight onto the cells, Organic Solar Cells, whereby the active material consists at least partially of organic dye, small, volatile organic molecules or polymer Amorphous silicon solar cell, Flexible amorphous silicon.

How the solar cell works:

We can recall that one silicon solar cell produces 0.5V to 0.6V; these cells are combines together to get a module. Usually 33/36 cells are connected together to make one module. This connection is made using the fact that the cells are connected electrically in series and/or parallel circuits to produce higher voltages, currents and power levels. One module has capability to produce enough voltage to charge 12 volt batteries and run pumps and motors. The PV module is basic building block of a PV power system. Most of the time PV modules are combined in series parallel manner to make PV array of required power complete power-generating unit.

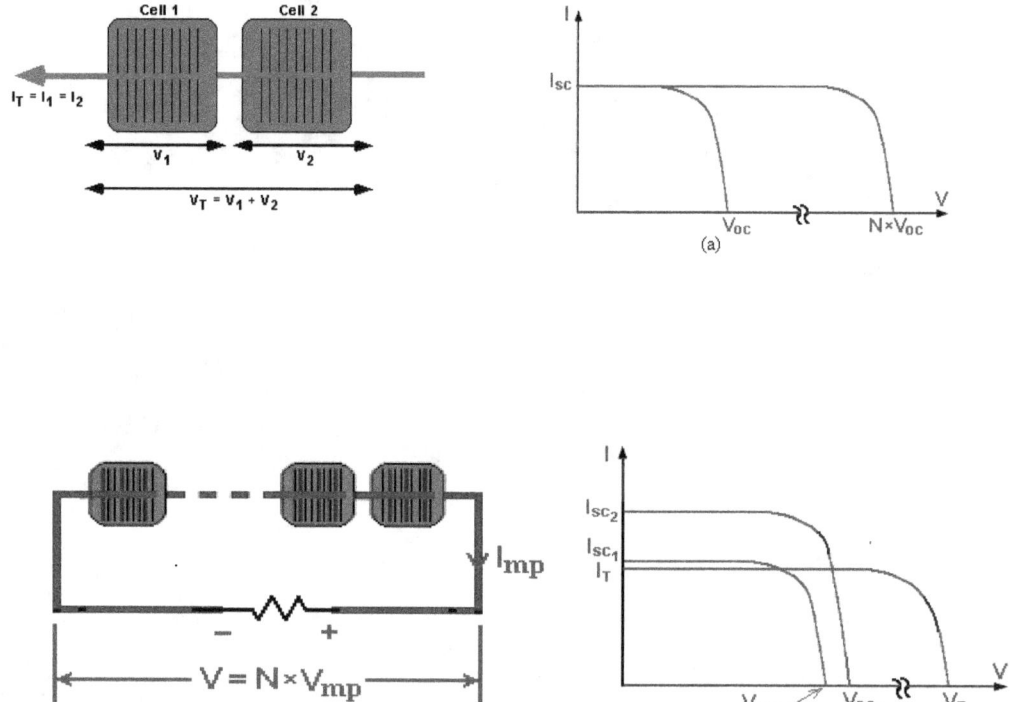

Picture: *Series connection of cells*

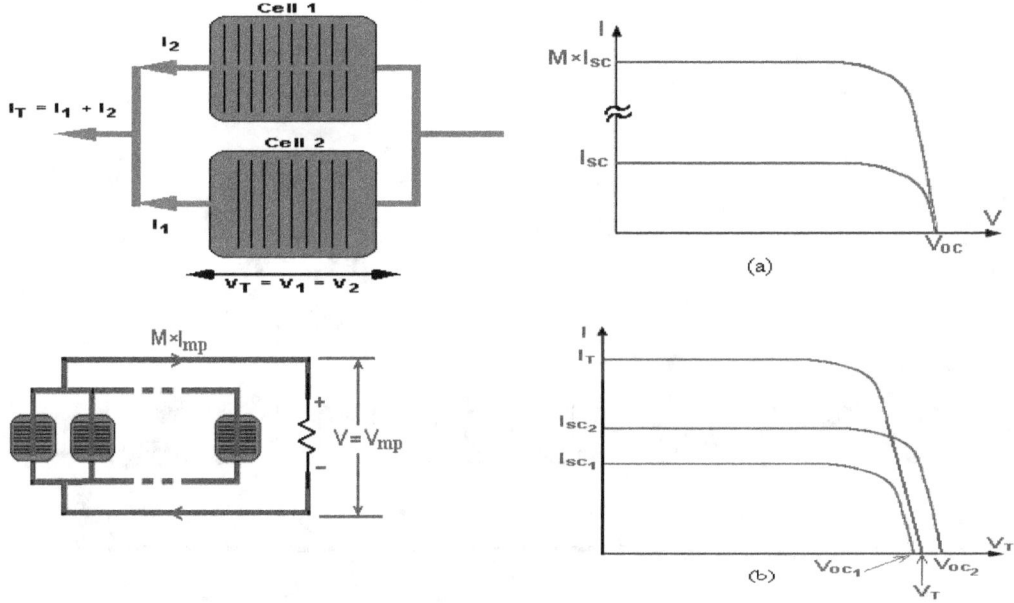

Picture: *Parallel connection of cells*

For two cells connected in series, the current through the two cells is equal. The sum of total voltage produced is the total of the individual cell voltages. As the current should same, a mismatch in current means that the total current from the configuration is equal to the lowest current Incase of parallel connected solar cell the voltage across the cell combination is always the same and the total current from the combination is the sum of the currents in the individual cells. Hot-spot heating occurs when there is one low current solar cell in a string of at least several high short-circuit current solar cells, as shown in the figure below:

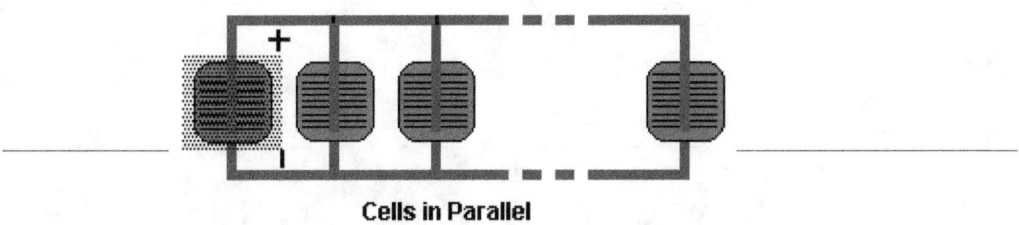

Cells in Parallel

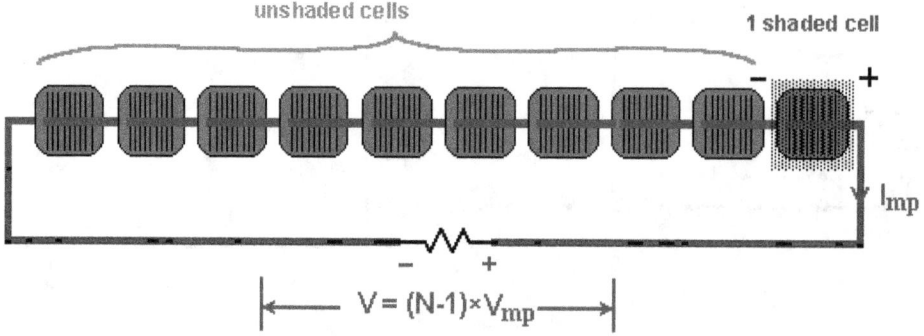

Picture: *Hot-spot*

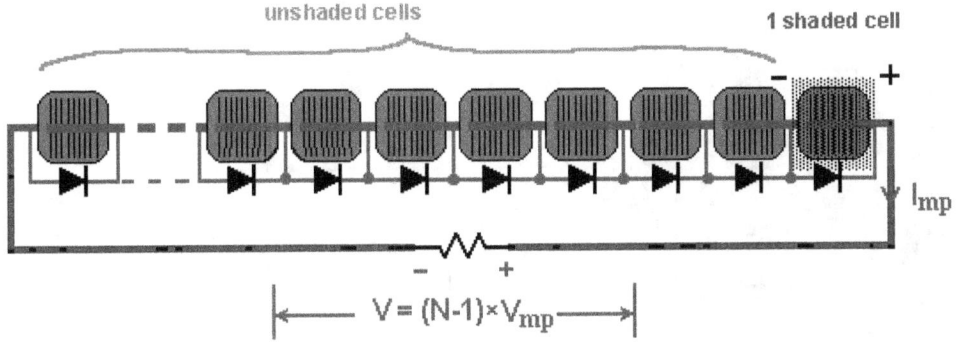

unshaded cells   1 shaded cell

I_{mp}

$$V = (N-1) \times V_{mp}$$

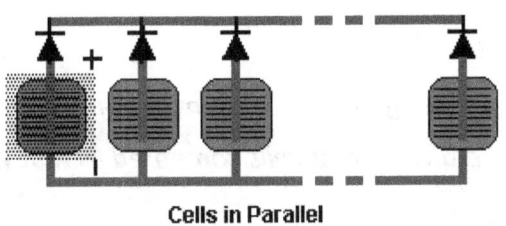

Cells in Parallel

.Picture: **Bypass diodes**

If all the solar cells in a module have identical electrical characteristics, and they all experience the same insolation and temperature, then all the cells will be operating at exactly the same current and voltage. In this case, the IV curve of the PV module has the same shape as that of the individual cells, except that the voltage and current are increased.

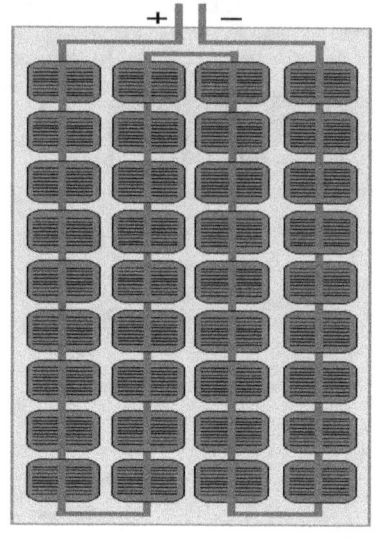

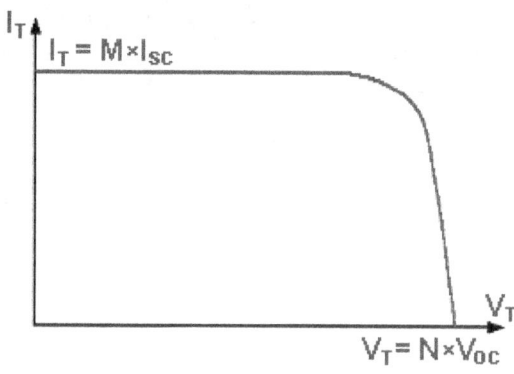

Picture: *PV module*

The packing density of solar cells in a PV module refers to the area of the module that is covered with solar cells compared to that which is blank. Packing density

Picture: *Packing Density*

The formula to measure the density of the packing density is as below:

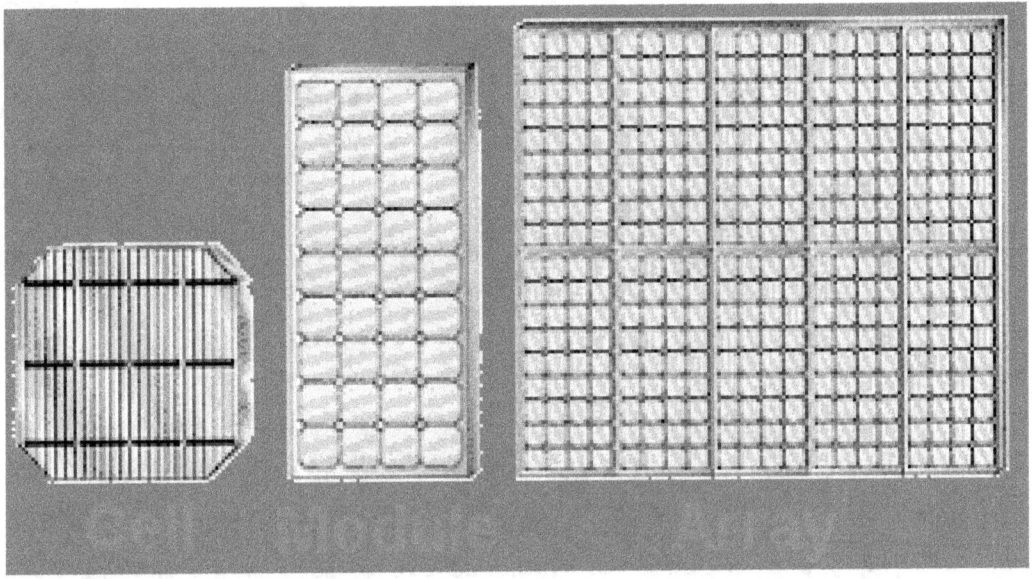

Picture: *Cell, Module and Array*

35

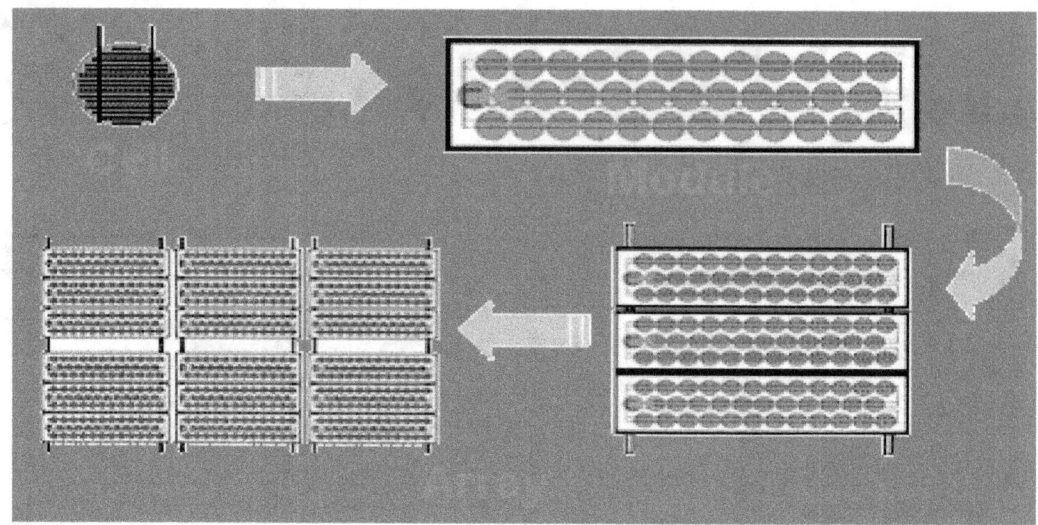

Picture: *Rotational picture of solar cell*

The performance of PV modules and arrays are generally rated according to their maximum DC power output (watts) under. Standard Test Conditions (STC). Standard Test Conditions are defined by a module (cell) operating temperature of 25o C (77 F), and incident solar irradiance level of 1000 W/m2 and under Air Mass 1.5 spectral distribution. Since these conditions are not always typical of how PV modules and arrays operate in the field, actual performance is usually 85 to 90 percent of the STC rating.

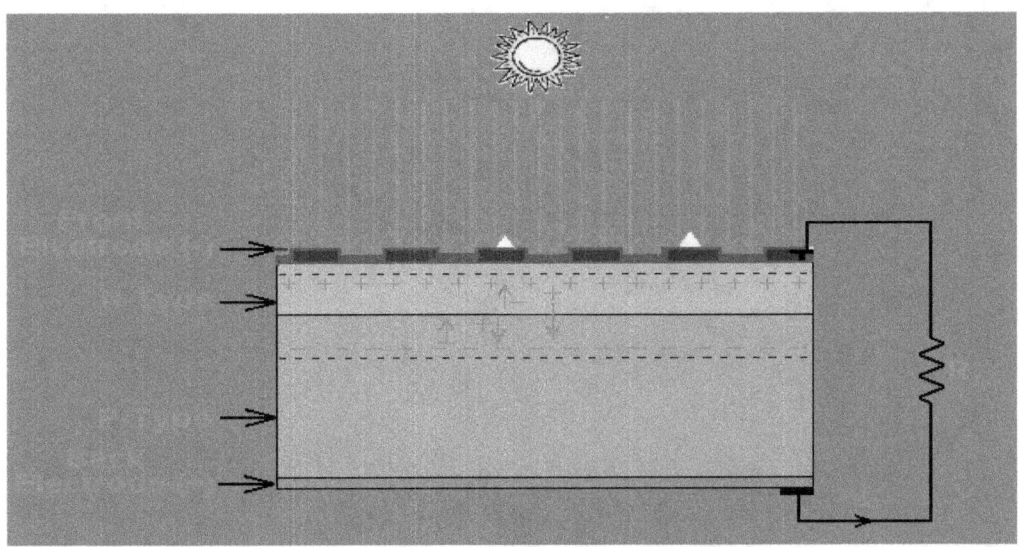

Picture: Diagram of photovoltaic cell

Today's photovoltaic modules are extremely safe and reliable products, with minimal failure rates and projected service lifetimes of 20 to 30 years. With the above PV modules the solar systems and electricity generating solutions are developed.

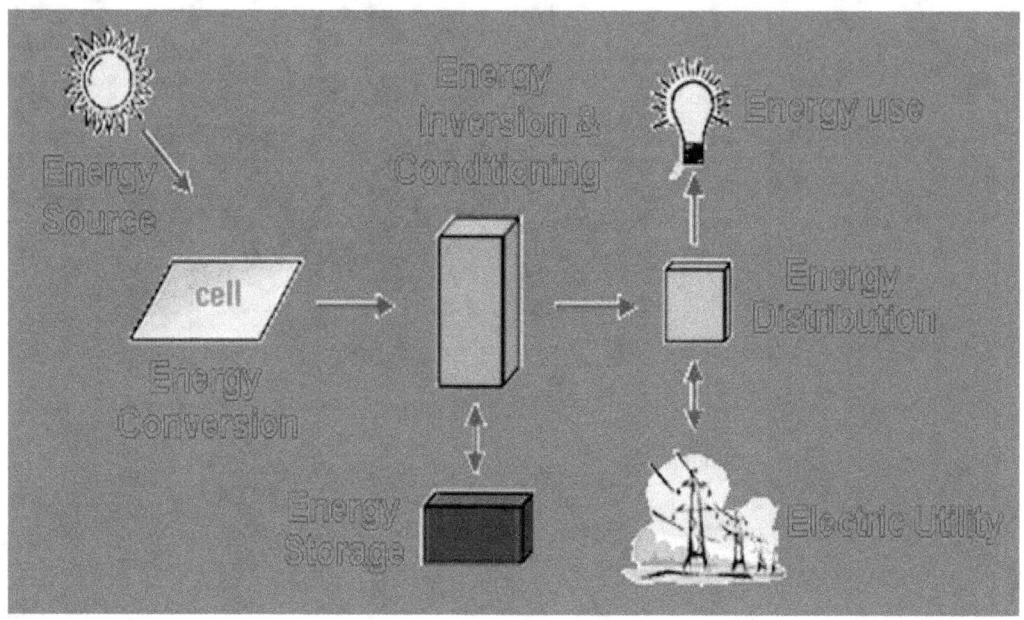

Picture: *Major Photovoltaic System Components*

PV systems are like any other electrical power generating systems. A PV array produces power by sunlight, a number of other components are required to properly conduct, control, convert, distribute, and store the energy produced by the array. Components required, and may include major components such as a DC-AC power inverter, battery bank, system and battery controller auxiliary energy sources and sometimes the specified electrical load (appliances). In addition, an assortment of balance of system (BOS) hardware, including wiring, over current, surge protection and disconnect devices, and other power processing equipment.

Different types of solar electric systems and how they can be built up:

These solutions can be different in category. Following is the category of solution those are possible to develop usually. They are:

1. Stand-alone PV systems: a) Simple single module DC system (PV Home system);b) Large DC PV system; c)PV system for DC and AC power

2. Hybrid PV systems: a) Hybrid-PV diesel; b) Hybrid PV-Diesel system for DC and AC power; c) Hybrid PV-Diesel system for DC and AC with generator for rectified DC d) Hybrid PV-Thermal systems; e) Hybrid PV-Wind systems

3. Grid-connected PV power systems: (a) Self-commutated; (b) Line commutated

> Simple single module DC PV systems (PV Home systems) are used for low cost rural electrification. Usually a single module is connected to the load for sunny day use.

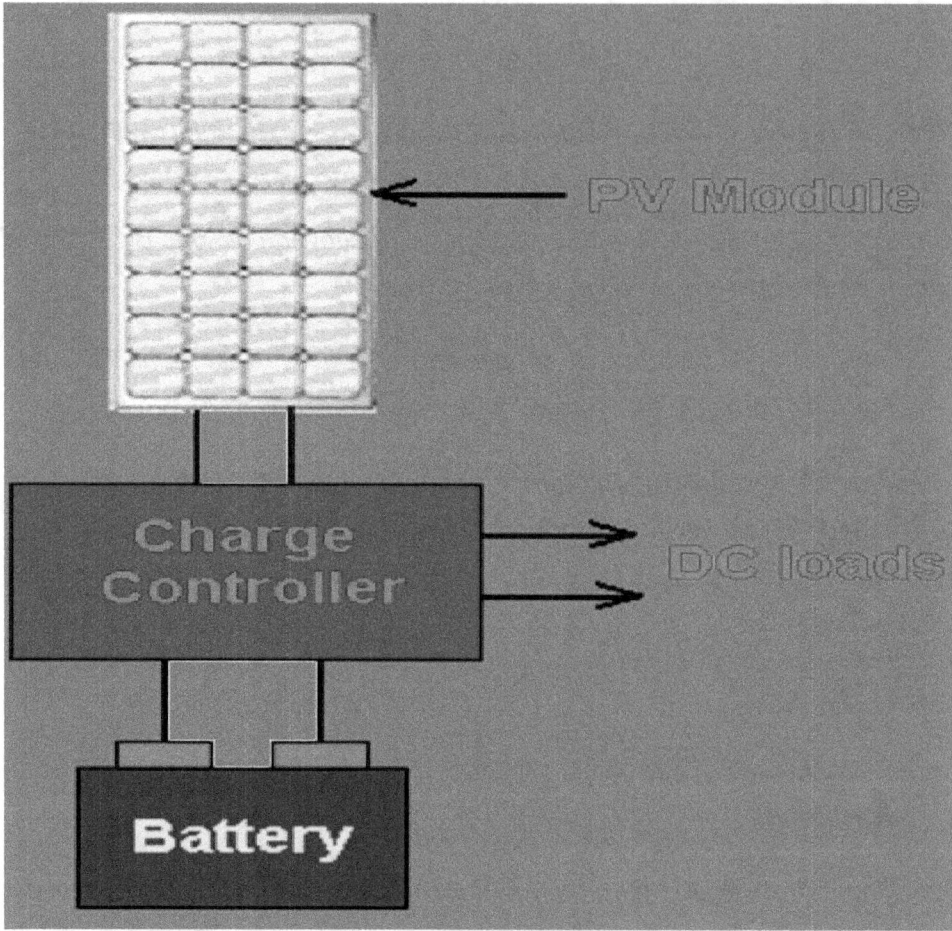

Picture: *Simple single module DC PV power system*

> *Simple single module DC PV system (PV Home system) is connected to a single low cost battery through a simple charge controller as the regulator that is connected to a CFL. The regulator may have a relay that could turn off the light if the battery voltage became too low. The user would have to wait until the module charged the battery back to an acceptable intermediate voltage before they could turn on the light again.*

> *Large DC PV power system consists of large number of modules and large number of batteries to drive heavy DC loads. A large heavy duty controller having high current driving capability is required. Usually Loads are connected through the charge controller via DC circuit breaker distribution box.*

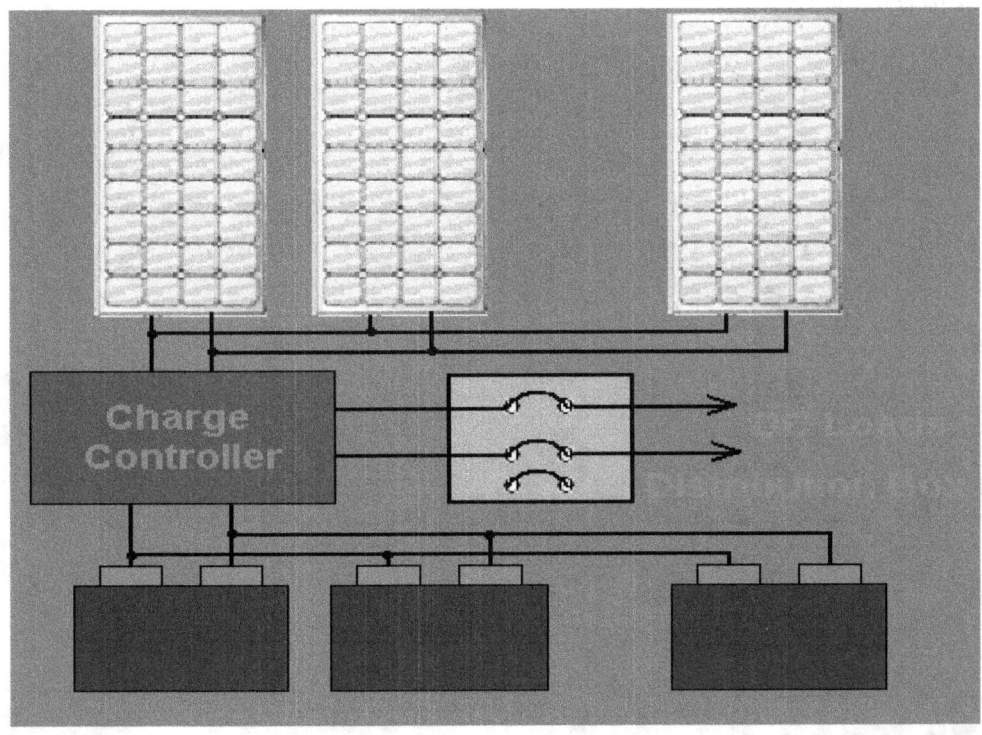

Picture: Large DC PV power system

> *PV system for DC and AC power consists of a large number of modules. Charge controller regulates the charging and discharging of the battery and the DC loads are supplied power through controller. Inverter generates AC power for the AC loads such as computers, fax machine, radio, TV, VCR and CD players.*

Hybrid PV-Diesel system

(i) Hybrid PV-Diesel system for DC and AC power:

Instead of relying purely on PV system for power, a system can be designed with other generators available as well. A common choice is a diesel generator. This generator produces AC power for AC loads which can be passed directly on to AC loads through transfer switches. Generator power can also be used after (converting AC to DC) rectification to charge batteries or to supply required DC input to the inverter.

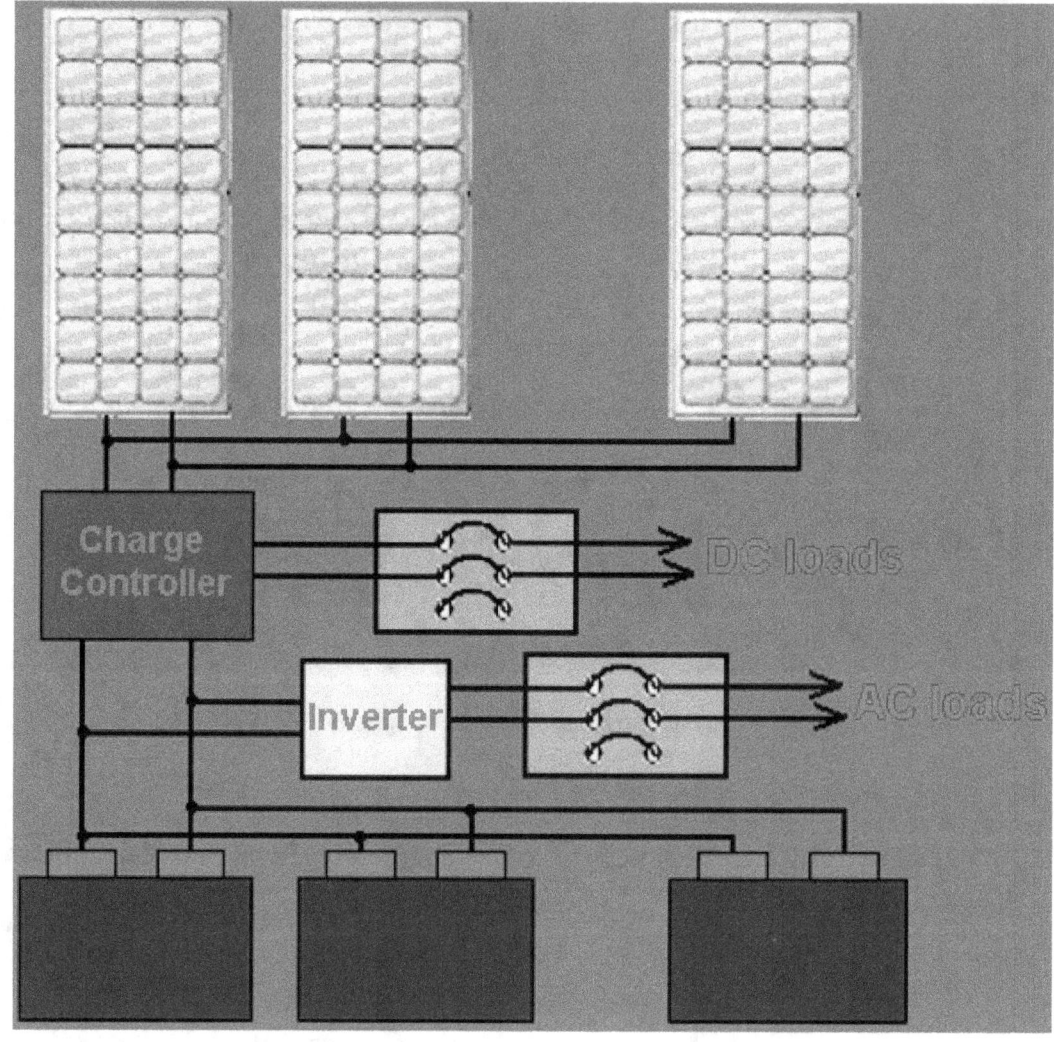

Picture: *AC and DC PV power system*

(ii) Hybrid PV-Diesel system for DC and AC with generator
  for rectified DC

A hybrid system can be designed to have the generator act only as a battery charger. AC power is not used to run the AC loads. All AC loads are driven by the power generated by the inverter only. The generator is turned on when the battery voltage is very low or weather is bad. After rectification generator output is used to charge the battery.

The generator needs to operate only for few hours to recharge batteries.

Generator is operated at its full rated output for maximum output and long life. When batteries are sufficiently recharged, the generator is turned off, and the finishing charge is supplied by the solar modules. If the bad weather continues, the generator is turned on again for a few days, and repeats the charging process. In this way the battery is kept fully charged having longer useful life.

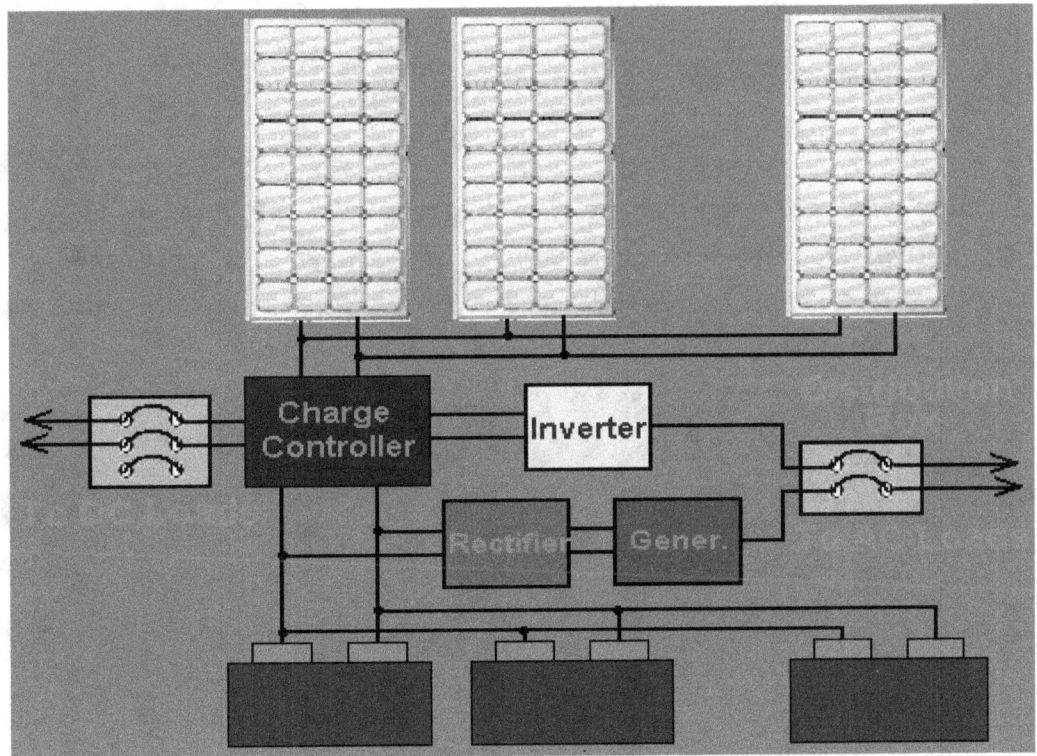

Picture: *Hybrid PV system with generator for AC and DC power*

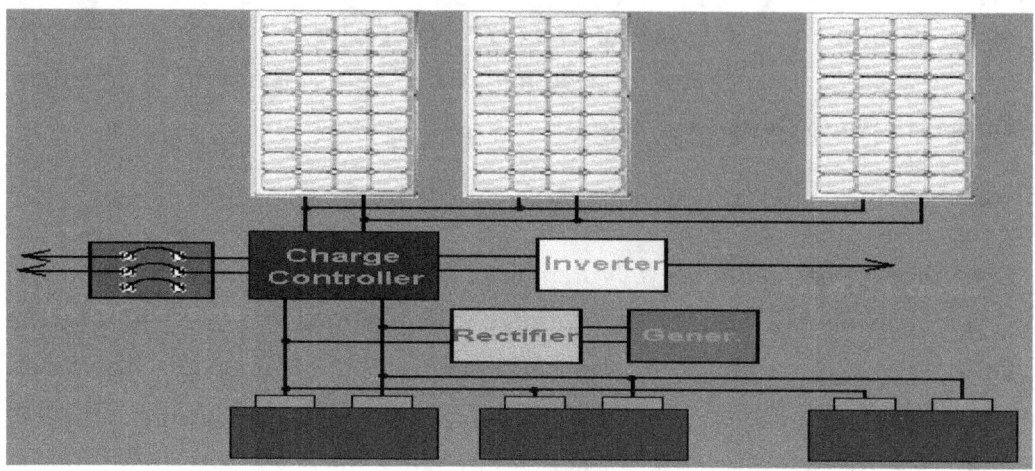

Picture: Hybrid PV system with generator for DC power

(iii) Hybrid PV-Thermal systems

Hybrid PV-thermal systems, *sometimes known as* **hybrid PV/T systems** *or* **PVT**, *are systems that convert solar radiation into electrical and thermal energy. These systems combine a photovoltaic cell, which converts electromagnetic radiation (photons) into electricity, with a solar thermal collector, which captures the remaining energy and removes waste heat from the PV module*

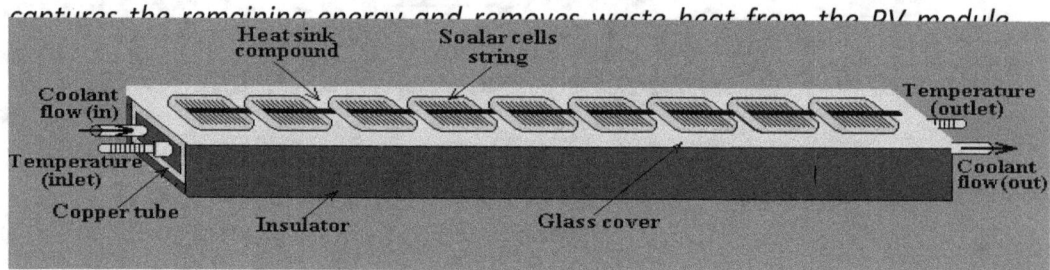

thus improving their efficiency by lowering resistance.

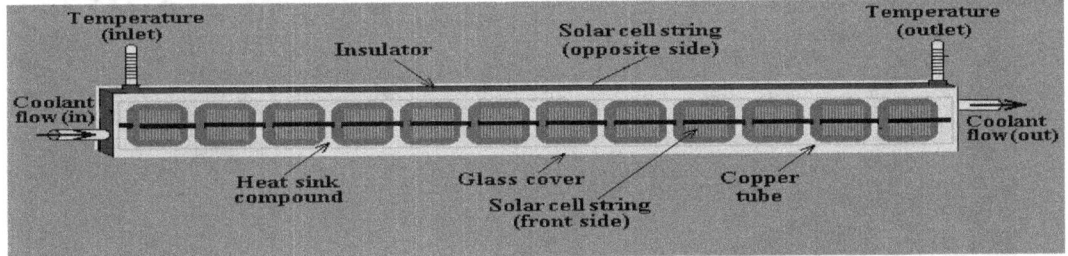

Picture: Solar cell string mounted on (a) flat horizontal and (b) flat vertical absorbers

Picture: Hybrid PV-Thermal systems installed at the roof-top of a building

Picture: One-axis tracking modular PV-Thermal concentrating collector for building as well as ground installations designed for hotel roof-tops and sports and recreations

Picture: Hybrid PV/Thermal system at the roof-top Roof from Conserval Engineering Inc. mounted on Chewonki Center for Environmental Education, Wiscasset, Maine, USA.

Hybrid PV-Wind systems

There are some places where wind speeds are often low in periods when the sun resources are at their best. On the other hand, the wind is often stronger in seasons when there are less sun resources. That can make solar PV-wind hybrid solutions an alternative to consider. Even during the same day, in many regions worldwide or in some periods of the year, there are different and opposite wind and solar resource patterns. And those different patterns can make the hybrid systems the best option for electricity production.

Hybrid PV-Wind system is the combination of PV and wind turbine for the generation of electricity.

Picture: *Hybrid PV-Wind systems*

Grid-connected PV systems

DC power generated by PV is converted into AC and is supplied to the national grid. Energy storage is not necessary in this case. On sunny days, the solar generator provides power, e.g., for the electrical appliances in the house. Excess energy is supplied to the national grid. During the night and overcast days, the house draws its power from the grid. In this way, the electricity grid can be regarded as a large "storage unit". In the case of a favorable rate-based tariff for PV electricity, as in force in some countries, it is more advantageous to feed all solar electricity into the grid.

Grid-connected PV systems can be subdivided into two kinds:

Decentralized Grid-connected PV systems: Decentralized Grid-connected PV systems have mostly a small power range and are installed on the roof-top of buildings (roof-top or flat-roof installation) or integrated into building facades. For example, in Germany around 80% of the more than 50,000 existing grid-connected PV systems are installed either on the roof-top of a building or integrated into a building façade. The benefit of the installation of a PV system into or onto a building is that no separate area for the solar generator is needed.

Centralized Grid-connected PV systems: Central Grid-connected PV systems have an installed power up to MW range. With such central photovoltaic power stations it is possible to feed directly into the medium or high voltage grid. Mostly central photovoltaic power stations are set on unused land, but in some cases an installation on buildings, mostly on the flat roof of greater buildings, is also possible.

Single phase less than or equal to 25kW. Three phase less than or equal to 300kW.

Grid-connected PV System can also be classified according to size:

Small – Power from 1 to 10 kWp. Typical applications are: rooftops of private houses, school buildings, car parks etc.

Medium size – Power from 10 kWp to some hundred of kWp. These kinds of systems can be found in what are called building integrated PV

(BIPV) systems, in roofs or facades. They may operate at higher voltages than smaller systems. Large size – Power from 500 kWp to MWp range, centralized systems.

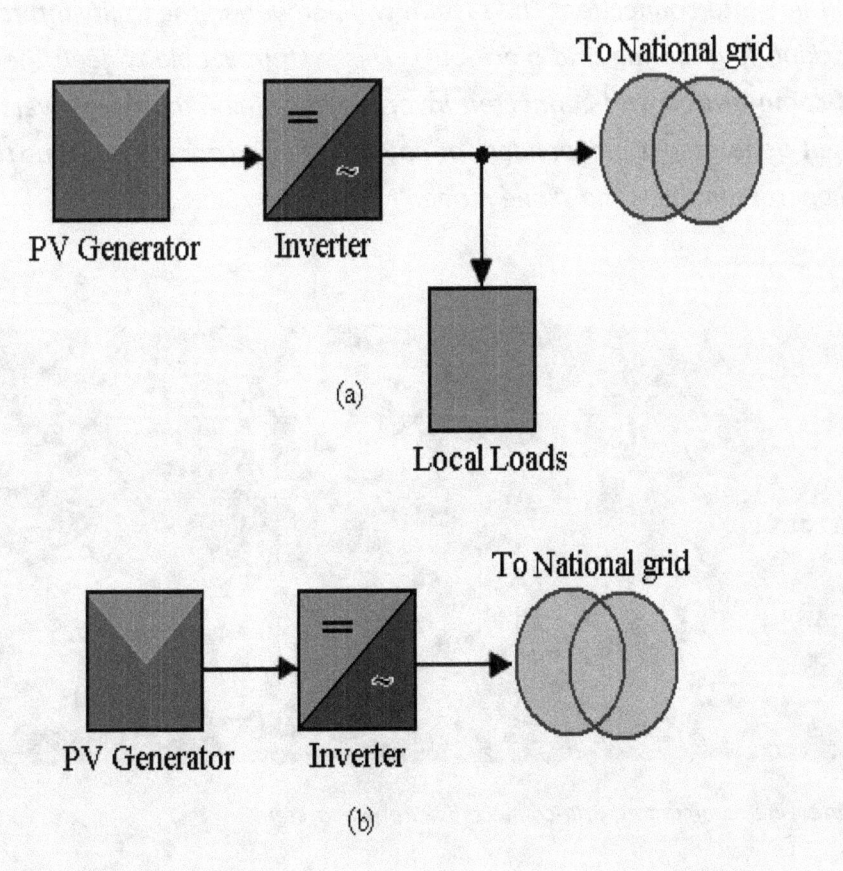

*These systems are normally operated by electric companies. In general a grid-connected PV system consists of the following two main components: **1) PV Module**; 2) **Inverter***

*The complete system consists of a support structure, cabling, and other conventional components. Besides the modules, the most important part of a system is the inverter. Inverters for Grid-connected PV power systems can be subdivided into the following classes; a) Line-commutated grid-connected inverter; b)Self-commutated grid-connected inverter. **Line-commutated grid-connected inverter:** In this inverter the AC output generated is dependent on the utility interconnection. The system will not work if the utility interconnection is lacking (e.g. during grid black-out). This system is able to feed the grid only. **Self-commutated grid-connected inverter:** In self-commutated inverter, the AC output generated is independent of any utility interconnection. Thus this system can operate in the stand-alone mode.*

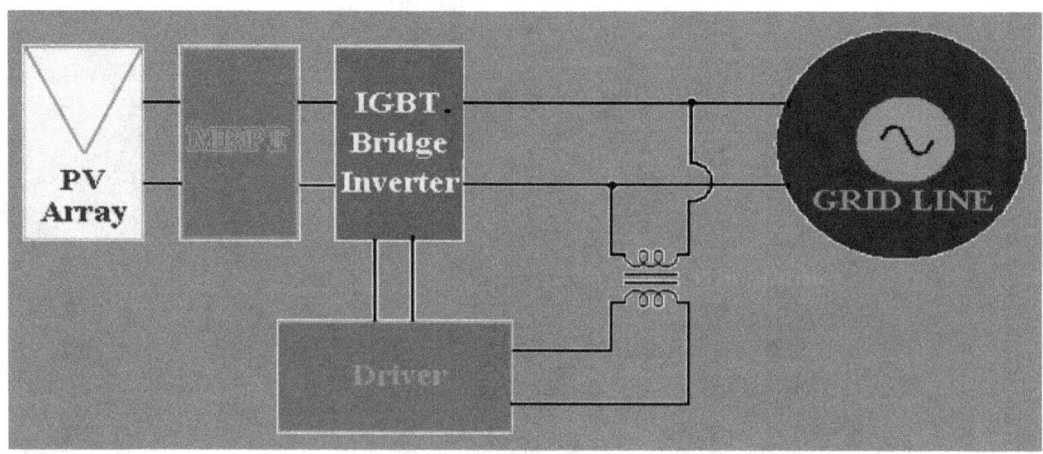

Picture: Line-commutated Grid-connected PV power system

Maximum Power Point Tracker (MPPT): *Maximum Power Point Tracking, frequently referred to as MPPT, is an electronic system that operates the Photovoltaic (PV) modules in a manner that allows the modules to produce all the power they are capable of. The current-voltage characteristics of photovoltaic module are such that it can be considered to be a constant current source up to a certain voltage range. This voltage is close to the open circuit voltage of the PV module. The I-V characteristic of a PV module as shown in, if a variable load resistance is connected across the solar array terminals, as shown in, and the load resistance is varied, the load lines corresponding to different load resistance intersect at different points on I-V curve resulting in different amount of power dissipation through RL. But the maximum would be dissipated if RL were adjusted to intersect the maximum power point (MPPT).*

Principle of Maximum power point tracking: *In a photovoltaic (PV) system it is desirable to extract the maximum amount of energy output from the PV array. This is possible if the array is operated at maximum power point (MPP) every instant, because the maximum power point is fluctuating due to change in isolation and temperature. With reference to, when the load resistance RL reaches RL2, the load line corresponding to RL2 intersects the maximum power point and maximum power will be dissipated through RL; we can then say that impedance matching between the PV system and the load is achieved. A photovoltaic array can deliver its maximum power only when the load impedance matches its dynamic impedance which however varies with insolation and temperature. Thus for maximum power transfer and minimization of mismatch losses, an impedance transformer is needed which will continuously match the dynamic impedance of the PV-array to the fixed impedance of the load for all insolation levels and temperature. Such an impedance matching device is called a maximum power point tracker (MPPT). It is seen that if the isolation is increased to I1 the MPP is shifted to P1 increasing the voltage V1 at maximum power point and when the isolation is decreased to*

I4 the MPP will be shifted to P4 decreasing the voltage V4 . Now it seems that if the PV-array is operated at the average value V0 of V1 and V4 it is quite possible that almost the maximum power P0 from the PV-array will always be supplied to the load. When this principle is adopted in designing an MPPT circuit and if the efficiency of the circuit is very high about 85-90% then approximately almost all the power can be delivered to the load. This type of MPPT is adequate for many practical purposes.

Maximum power point tracking can also be conveniently achieved by a pulse width modulated DC to DC converter having a variable duty cycle which is a function of the light intensity. The duty cycle can be varied by automatic control circuit. Automatic duty cycle adjustment can be achieved by using simple circuit or by a microprocessor based control circuit. The microprocessor based circuit hunts for the maximum power point more accurately under all circumstances (insolation and temperature). On the other hand simple control circuit tracks the maximum power point approximately rather than accurately.

Step-down or Buck type MPPT: A step down or buck type MPPT. This type of circuits are, in fact, current boosters and are generally employed to drive current receiving loads such as permanent magnet DC motor loaded with a constant torque mechanical load e.g. volumetric pump. In this circuit, the output voltage is always less than the input voltage and can be practically anywhere from 10% to 90% of the input voltage. This means that this circuit can be used as a DC step down transformer with highest efficiency. In the figure the output voltage is compared with a stable reference voltage and the amplified error signal is used to generate a pulse width modulated waveform, which controls the on-off periods of the switching transistor Tr. The current flows through the inductor, into the output capacitor and to the load when the switch is turned on. When the output voltage exceeds the reference voltage, the switching transistor Tr is turned off. At this instant the stored energy in the inductor reverses its polarity, takes the path through the diode and sends the current into the loads while the voltage is maintained by the capacitor. When all the stored energy in the inductor is used up the capacitor discharges and the output voltage decreases. At this step, the switch is turned on and the process continues such that the output voltage is maintained very close to the reference voltage.

Step-up or boost type MPPT: *This MPPT circuit is actually a voltage booster circuit. This circuit is generally employed for battery charging application and show a boost type regulator. In this circuit the conduction of the switching transistor Tr is controlled by the pulse width modulated signal. When the switching transistor Tr is turned on, the current flows through the inductor and energy is stored in it. When the switch is turned off, the stored energy in the inductor tends to collapse and its polarity changes such that it adds to the input voltage. Thus the voltage across the inductor and the input voltage are in series and together charge the output capacitor to a voltage higher than the input voltage.*

Charge Controllers in PV Systems: *The primary function of a charge controller in a stand-alone PV system is to maintain the battery at highest possible state of charge while protecting it from overcharge by the array and from over-discharge by the loads. Without charge control, the current from the array will flow into a battery proportional to the irradiance, whether the battery needs charging or not. If the battery is fully charged, unregulated charging will cause the battery voltage to reach exceedingly high levels, causing severe gassing, electrolyte loss, internal heating and accelerated grid corrosion. In most cases if a battery is not protected from overcharge in PV system, premature failure of the battery and loss of load are likely to occur. When a battery is over-discharged, the reaction in the battery occurs close to the grids, and weakens the bond between the active materials and the grids. When a battery is repeatedly over-discharged, a loss of battery capacity and life will eventually occur.*

Important functions of a charge controller are:

*a) **Prevent Battery Over-charge:** to limit the energy supplied to the battery by the PV array when the battery becomes fully charged.*

*b) **Prevent Battery Over-discharge:** to disconnect the battery from electrical loads when the battery reaches low state of charge.*

*c) **Provide Load Control Functions:** to automatically connect and disconnect an electrical load at a specified time, for example operating lighting load from sunset to sunrise. In order to obtain a long battery life time, over-charging and over-discharging must be avoided. In the case of lead-acid battery the voltage is a measure of its state-of-charge. Therefore, by measuring this voltage, one can determine whether the battery is working outside its normal regime.*

__Charge controllers set points:__ The battery voltage levels at which a charge controller performs switching functions is called the charge controllers set points. There are four basic switching set points which are defined for most charge controllers that have battery over-charge and over-discharge protection features.

These are: a) Voltage Regulation (VR) set point;

b) Array Reconnect Voltage (ARV) set point;

c) Low voltage Load Disconnect (LVD) set point;

d) Load Reconnect Voltage (LRV) set point.

Voltage Regulation (VR) set point: *The voltage regulation set point is defined as the maximum voltage that the charge controller allows the battery to reach, limiting the overcharge of the battery. Once the controller senses that the battery reaches the voltage regulation set point, the controller will either discontinue battery charging or begin to regulate (limit) the amount of current delivered to the battery.*

Array Reconnect Voltage (ARV) set point: *When the battery voltage decreases to a predefined voltage, the array is again reconnected to the battery to resume charging. This voltage at which the array is reconnected is defined as the array reconnect voltage (ARV) set point.*

Voltage Regulation Hysteresis (VRH): *The voltage difference between the voltage regulation set point and the array reconnect voltage is often called the voltage regulation hysteresis (VRH). If the hysteresis is too great, the array current remains disconnected for long periods, effectively lowering the array energy utilization and making it very difficult to fully recharge the battery. If the regulation hysteresis is too small, the array will cycle on and off rapidly, perhaps damaging controllers which use electro-mechanical switching elements.*

Low Voltage Load Disconnect (LVD) Set Point: *If battery voltage drops too low, due to prolonged bad weather for example, certain non-essential loads can be disconnected from the battery to prevent further discharge. This can be done using a low voltage load disconnect (LVD) device connected between the battery and non-essential loads. The LVD is either a relay or a solid-state switch that interrupts the current from the battery to the load, and is included as part of most controller designs. In some cases, the low voltage load disconnect unit may be a separate unit from the main charge controller.*

Load Reconnect Voltage (LRV) Set Point: *The battery voltage at which a charge controller allows the load to be reconnected to the battery is called the load reconnect voltage (LRV). After the charge controller disconnects the load from the battery at the LVD set point, the battery voltage rises to its open-circuit voltage. When additional charge is provided by the array, the battery voltage rises even more. At some point, the controller senses that the battery voltage and state of charge are high enough to reconnect the load, called the load reconnect voltage set point.*

Low Voltage Load Disconnect Hysteresis (LVDH): *The voltage difference between the LVD set point and the load reconnect voltage is called the low voltage disconnect hysteresis (LVDH). If the LVDH is too small, the load may cycle on and off rapidly at low battery state-of-charge (SOC), possibly damaging the load or controller, and extending the time it takes to fully charge the battery. If the LVDH is too large, the load may remain off for extended periods until the array fully recharges the battery. With a large LVDH, battery health may be improved due to reduced battery cycling, but with a reduction in load availability.*

These are typical, presented here only for example. *Typical set points for 12 V lead-acid batteries at 77°F (25°C); Voltage Regulation set point (VR): 14.4 V; Array Reconnect Voltage set point (ARV):13.0 V; Load Reconnect Voltage set-point (LRV):12.5 V. Low Voltage load Disconnect (LVD):10.8 V. Temperature compensation for 12V battery : 0.03 V per °C deviation from standard 25°C.*

Charge controller design: *Charge regulation is the primary function of a charge controller, and perhaps the single most important issue related to battery performance and life. The regulation circuit in a PV system continuously* monitors and compares the state-of-charge of the battery with predetermined reference value and takes decision regarding turning on and off of the battery and charging or trickle charging the battery. There are three types of regulations: a) Self-regulation; b) Shunt regulation; c) Series regulation.

Self-regulation: The principle of self-regulation is to size the photovoltaic generator so that its voltage sensitive region coincides with the battery critical region i.e. 90-95% state-of-charges. Let us suppose that a 12V lead-acid battery has a charge voltage that ranges from 12.8V at 60% depth of discharge to 14.4V at full charge. The voltage operating point of the array that would transfer maximum power from the array is 14.4V plus the voltage drop across blocking diode or a total of (14.4+0.7) = 15.1V. The PV array is thus maintained at open circuit by reverse biasing the diode until the battery voltage falls and charging is needed once again.

Shunt regulation: Since photovoltaic cells are current-limited by design (unlike batteries), PV modules and arrays can be short-circuited without any harm. The ability to short-circuit modules or an array is the basis of operation for shunt controllers.

An electrical design of a typical shunt type controller: The shunt controller regulates the charging of a battery from the PV array by short-circuiting the array internal to the controller. All shunt controllers must have a blocking diode in series between the battery and the shunt element to prevent the battery from short-circuiting when the array is regulating. Because there is some voltage drop between the array and controller and due to wiring and resistance of the shunt element, the array is never entirely short-circuited, resulting in some power dissipation within the controller. For this reason, most shunt controllers require a heat sink to dissipate power, and are generally limited to use in PV systems with array currents less than 20 amps.

Series regulation: *Series regulator works in series between the array and battery. They use some type of control or regulation element in series between the array and the battery. While this type of controller is commonly used in small PV systems, it is also the practical choice for larger systems due to the current limitations of shunt controllers.*

A typical series type controller: *In a series controller, a relay or solid-state switch either opens the circuit between the array and the battery to discontinuing charging, or limits the current in a series-linear manner to hold the battery voltage at a high value. In the simpler series circuit design, the controller reconnects the array to the battery once the battery falls to the array reconnect voltage set point. As these on-off charge cycles continue, the 'on' time becoming shorter and shorter as the battery becomes fully charged. Because the series controller open-circuits rather than short-circuits the array as in shunt-controllers, no blocking diode is needed to prevent the battery from short-circuiting when the controller regulates.*

Storage System: *Another important component of the solar system is the storage system. Electric power is generated only during day time by PV panels. But we need light at night. So some sort of storage systems required to use with PV system. Earlier I have discussed on the necessity of using alternative clean and green energy such as solar photo voltaic (PV) panels for the resources of clean and green energy. Fossil fuel and other resources of power and electricity has limit and will be finished off with in near future. In early version I have discussed on PV panels and how they work. Not only in my early article; there are many readings and books are available to take knowledge on solar PV panels in market now a days. Many analyst and expert had discussed on solar PV panels in different magazine and news paper in recent time.*

There are some more components and equipments are using in a solar powered electric system. Those equipments are very important role to play in a solar powered system. I think people should know about these for their knowledge and usability.

*One important component in the PV panels is the storage system. It is required to have a storage system in a solar powered electricity generating system. The major role of this storage system is to store the electricity and release it whenever required for. As we know Electric power is generated only during day time by PV panels. But we need light at night. So some sort of storage systems has to be used with PV system. Although various types of batteries are available in the market, lead-Acid batteries are widely used in PV systems. Many of us may know that In **1859** French physicist Gaston Planté invented this rechargeable battery. Despite having a very low energy-to-weight ratio and a low energy-to-volume ratio, they have the ability to supply high surge current and relatively large power-to-weight ratio. These features, along with their low cost, make them attractive for use in motor vehicles and now in PV systems. Their huge Ampere-Hour capacity makes them suitable for PV system.*

The lead acid batteries may be divided in to some parts.

Parts of a Lead-acid Battery:

*A battery consists of a number of cells and each cell of the battery consists of **(1) positive and negative plates (2) separators (3) electrolyte** and **(4)** the container. Different parts of a lead-acid battery are described below:*

(1) Plates. *A plate consists of a lattice type of grid of cast antimonied lead alloy which is covered with active material. The grid not only serves as a support for* the fragile active material but also conducts electric current. Grids for the positive and negative plates are often of the same design although negative-plate grids are made somewhat lighter.

(2) Separators. These are thin sheets of a porous material placed between the positive and negative plates for preventing contact between them and thus avoiding internal short-circuiting of the battery. A separator must, however, be sufficiently porous to allow diffusion or circulation of electrolyte between the plates. These are made of especially-treated **cedar wood, glass wool mat, micro porous rubber (mipor), micro porous plastics (plastipore, miplast)** and **perforated p.v.c.** In addition to good porosity, a separator must possess high electrical resistance and mechanical strength.

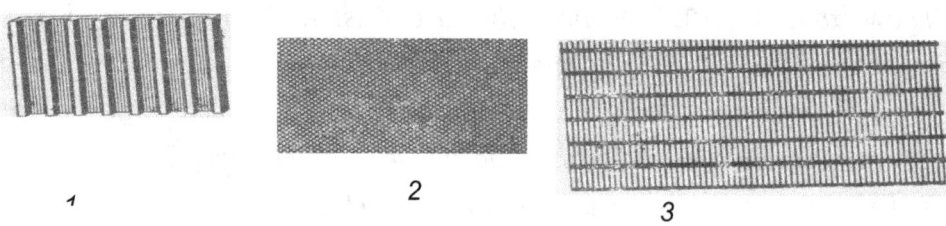

1 2 3

Separators: (1) and (2) Miplast type (3) Perforated type

(3) Electrolyte. It is dilute sulphuric acid which fills the cell compartment to immerse the plates completely.

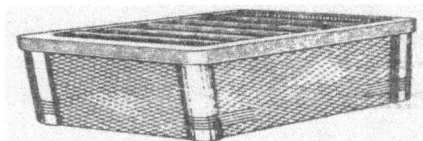

Battery container

Followings are the parts of the lead acid battery picture shown as parts separately.

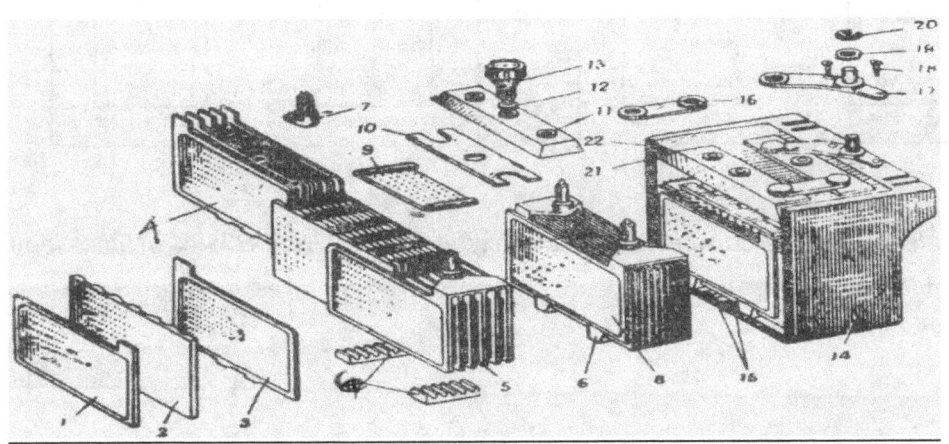

1. Negative plate 2. Separator 3. Positive plate. 4. Positive group 5. Negative group 6. Negative group grooved support block 7. Lug 8. Plate group 9. Guard screen 10. Guard plate 11. Cell cover 12. Plug washer 13. Vent plug 14. Monoblock jar 16. Supporting prisms of + ve group 16. Inter-cell connector 17. Terminal lug 18. Screw 19. Washer 20. nut 21. Rubber packing 22. Sealing compound.

Discharging:

When the cell is fully charged, its positive plate or anode is PbO_2 and the negative plate or cathode is Pb. When the cell discharges i.e. it sends current through the external load, then H_2SO_4 is dissociated into positive H_2 and negative S04 ions. As the current within the cell is flowing from cathode to anode, H2 ions move to anode and SO4 ions move to the cathode.

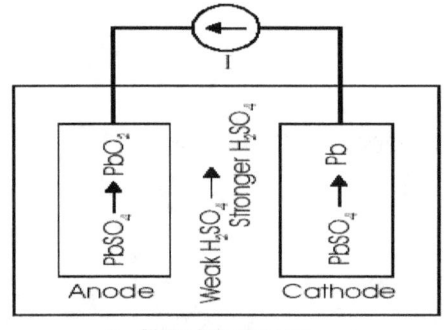 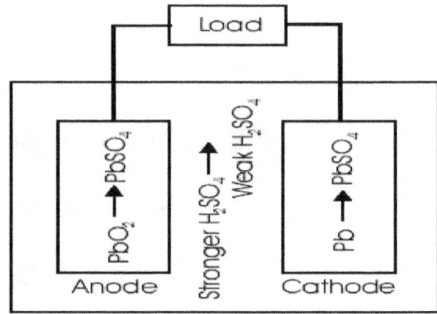

a. Charging process   b. Discharging process

At anode (Pb02), H2 combines with the oxygen of Pb02and H2SO4 attacks lead to form PbSO4.

CHARGING

When the cell is recharged, then H2 ions move to cathode and S04 ions go to anode and the following changes take place:

$$PbSO_4 + 2H_2O \rightarrow PbO_2 + 4H^+ + SO_4^{2-} + 2e^- \qquad (3.7a)$$

and

$$PbSO_4 + 2e^- \rightarrow Pb + SO_4^{2-}. \qquad (3.7b)$$

Hence, the anode and cathode again become PbO2 and Pb respectively.

It will be noticed that during charging:

The anode becomes dark chocolate brown in color *(PbO2) and* cathode becomes grey metallic lead (Pb). Due to consumption of water, specific gravity of H2SO4 is increased; there is a rise in voltage, energy is absorbed by the cell.

Efficiency:

Ideally, the charging and discharging processes of the lead-acid system should be reversible. In reality, however, they are not. Some of the electrical energy intended for charging is lost in the internal resistance and is converted to heat. When hydrogen is lost, it also represents an energy loss. Typically, the charging process is about 95% efficient. The discharge process also results in some losses due to internal resistance of the battery, so only about 95% of the stored energy can be recovered. The overall efficiency of charging and discharging a lead-acid battery is thus about 90%.

*Since battery losses to internal resistance are proportional to the square of the current, this means that **high current charging or high current discharging** will tend to result in higher internal losses and **less overall performance efficiency**.*

Amount of stored energy:

*The amount of energy stored in a battery is commonly measured in **ampere hours**. While ampere hours are technically not units of energy, but, rather, units of charge, the amount of charge in a battery is approximately proportional to the energy stored in the battery. If the battery voltage remains constant, then the energy stored is simply the product of the charge and the voltage.*

Effect of DOD on Life-Cycle:

The following Figure shows how the depth of discharge affects the number of operating cycles of a deep discharge battery. The PV system designer must carefully consider the trade-off between using more batteries operating at shallower discharge rates to extend the overall life of the batteries vs. using fewer batteries with deeper discharge rates and the correspondingly lower initial cost.

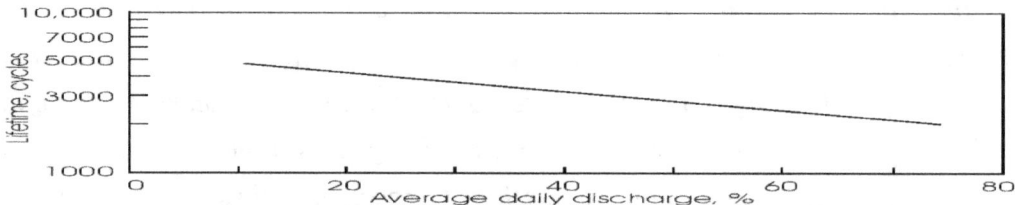

Lead-acid battery lifetime in cycles vs. depth of discharge per cycle

Vented and nonvented batteries:

In certain lead-calcium batteries, minimal hydrogen and oxygen are lost during charging. This means minimal water is lost from the electrolyte. As a result, it is possible to seal off the cells of these batteries, making them essentially maintenance free. The trade-off, however, is that if these batteries are either purposely or inadvertently discharged to less than 75% of their maximum charge rating, their expected lifetime may be significantly shortened.

Lead-antimony electrodes, on the other hand, may be discharged to 20% of their maximum rating. This means that a 100 Ah lead-calcium battery has only 25 Ah available for use, while a 100 Ah lead-antimony battery has 80 Ah available for use, or more than 3 times the availability of the lead-calcium unit.

However, the lead-antimony unit produces significantly more hydrogen and oxygen gas from dissociation of water in the electrolyte, and thus water must be added to the battery relatively often to prevent the electrolyte level from falling below the top of the electrodes. Water loss can be reduced somewhat by the use of cell caps that catalyze the recombination of hydrogen and oxygen back into water, which returns to the cell.

Chemistry of the Nickel Cadmium Storage Battery:

Ni-Cd batteries use **nickel oxi-hydroxide** for the anode plates and **finely divided cadmium** for the cathode plates.

The electrolyte in the Ni-Cd system is potassium hydroxide (KOH).

The NiOOH anode is generally made of nickel fibers mixed with graphite- or nickel-coated plastic fibers. Small quantities of other materials such as barium and cobalt compounds are also added to improve performance. The cathode is also frequently made of a cadmium-coated plastic fiber. If the cathode is not a coated plastic, then it is commonly mixed with iron or nickel. The fiber structures of anode and cathode maximize the surface area while minimizing the amount of relatively expensive nickel and cadmium required for the electrodes.

On discharge, the NiOOH of the positive plate is converted to Ni(OH)2 and the cadmium metal of the negative plate is converted to Cd(OH)2.

The basic reactions are:

At the positive plate: **NiOOH + H2O + e- = Ni(OH)2+OH**

At the negative plate: **Cd + 2OH- = Cd(OH)2 + 2e-**

Overall: **2NiOOH + Cd + 2H20 = 2Ni(OH)+Cd(OH)**

The voltage of the fully charged cell is 1.29 V. Unlike the lead-acid system where the specific gravity of the electrolyte changes measurably during discharge or charge, the KOH electrolyte of the Ni-Cd system changes very little during battery operation.

Ni-MH: (Nickel Metal Hydride)

Another technology that is becoming very popular, particularly in smaller applications such as camcorders and laptop computers, is the **nickel-metal hydride** (NiMH) battery. This battery replaces the cadmium cathode with an environmentally benign **metal hydride cathode**, allowing for higher energy density at the cathode and a correspondingly longer lifetime or higher capacity, depending on the design goal. The anode is the same as in the Ni-Cd cell and KOH is used as the electrolyte. The overall discharge reaction is

$$MH + NiOOH = M + Ni(OH)2$$

Lithium Ion (Li-Ion) Battery:

Lithium-Ion battery is an advanced battery. Lithium ion battery has high energy density, low self discharge and **no memory effect**. These batteries are used where smaller and light weight batteries are required such as cell phones and laptop computers.

However the main disadvantages of these batteries are the **high price** and **fewer charge/discharge** cycles

Electric Double Layer Capacitor (Super Capacitor):

EDLC is produced by placing appropriate electrodes in a suitable electrolyte. When charging voltage is applied as shown in Fig. 3.2, the boundaries of the two electrodes and electrolyte forms a cell with two capacitors in series. By charging the cell, the ions in the electrolyte are attracted to the surface of each electrode, which face the charges on the electrodes forming electric double layer in between. The charging energy remains stored in the double layers as electrostatic energy. It's time to discuss about some more alternative and renewable energy. Among them the most popular is SOLAR COOKER DRYER. In other word we can say thermal application of the solar energy. The radiation of sun can be converted in to thermal applications by which the sun shine can be converted in to heat and cook the necessary food or things. In any collection device, the principle usually followed is to expose a dark surface to solar radiation so that the radiation is absorbed: i) Water heating; ii) Space heating; iii) Power generation; iv) Space cooling and refrigeration; v) Distillation; vi) Drying; vii) Cooking.

Solar Cooker: The temperature required for cooking of many foods is about 1000C, but the temperature of the flame is quite high, resulting in heat transfer. Generally, where electricity or gas is used for cooking, the burner is rated at 1.0kW which brings about 2 liters of water to boiling point in about 10 minutes. Therefore a solar cooker should be designed such that it provides about 1.0kW of energy which can be obtained with 2.0m2 flat plate collector with efficiency of 50%. Several basic types of solar cookers have been developed to date. These cookers are broadly divided into three types- 1) Direct or focusing type: In these cookers some kind of a solar energy concentrator is used which when directed towards the sun focuses the solar radiation on a focal point or area on which a cooking or frying pan is placed. In these cookers the convection heat loss from cooking vessel in large and the cooker utilizes only the direct solar radiation. It is observed that under clear and calm days, one litre of water comes to boiling point within 25 minutes. Variety of foods can be cooked with this cooker and its output is equivalent to that of and electric hot plate of about 400watts. This cooks rice, pulses, meat and other food within half an hour in good sunshine. Use of pressure cooker/vessels of up to 13 liters capacity, aluminum reflectors, needs tracking, temperature>2000C, about 2m diameter; 2) Indirect or box type: In these solar cookers an insulated hot box (square, rectangular, cylindrical) painted black from inside with double glazing is used. To enhance the solar radiation, plane sheet reflectors (single or multiple) are used. Here, the adjustment of cooker towards the sun is not so frequently required as in the case of direct type solar cooker. This is a slow cooker and takes long time for cooking and many of the dishes cannot be prepared with this cooker. Box type solar cooker consists of a rectangular enclosure insulated on the bottom and sides and having one or two glass covers on the top. Solar radiation enters through the top and heats up the enclosure in which the food to be cooked is placed in shallow vessels. In this cooker, as expected, the cooking is slow and the time required for cooking rice and chopped meat varies between 2 and 3 hours depending on the sun's intensity. Typical size of insulated box1m×1m×0.2m single or double glass covers with one two boosters black paint or coating 3 or 4containers1200C, rice/ pulses/ meat. Baking difficult is a problem regarding this cooker but once any person can overcome the problem; can easily use this cooker when ever needed. 3) Advanced type: The problem of cooking outdoors is avoided to some extent. The cooking in some cases can either be done with stored heat or the solar heat is directly transferred to the cooking vessel in the kitchen. The cookers use either a flat-plate or focusing collector which collect the solar heat and transfer this to the cooking vessel. Heat transfer type solar cooker consists of two parts - a cylindro plarabolic reflector (3m×2m×0.5m); - a hot box. (0.4m×0.4m ×1.2m). The tube contains oil which when heated rises due to natural convection to the insulated hot box. For automatic tracking, a simple clock mechanism can be used. The temperature at the top of the reservoir on sunny days reaches 1500C and at night rarely falls below 1000C. All types of cooking except frying can be done with this cooker. Parabolic reflector with point focus on the ground inside the kitchen: 7m2 reflector area; Continuous diurnal tracking; Temperature>1100C; Cooking for 10-15 persons.

Performance of Box type solar cooker: The performance of a simple box type solar cooker depends on many parameters such as climatic parameters like solar radiation, ambient temperature, wind speed etc.; design parameters like properties of black paint used in tray, type of cooker, insulation properties, effectiveness of reflector-booster system etc and on operational parameters like amount and type of food kept for cooking, number of cooking vessels used etc. Generally as a matter of routine, three parameters are experimentally measured: 1) temperature-time curve of solar cooker when it is empty 2) cooking times of different food products and 3) the time required to heat a known amount of water up to the boiling point. A little design variation in solar cooker will affect the above three parameters.

Testing of a solar cooker: Generally cookers are thermally rated according to 1) stagnation plate temperature 2) time required for cooking different food products and 3) time required to bring a known amount of water to the boiling point. The above procedures are not acceptable universally since the testing will depend on the climatic parameters. For evaluating the hot box type solar cooker two figure of merits, F1 and F2 can be determined which are independent of the climatic parameters. These figures can be found by conducting the stagnation temperature test (without load) and by sensibly heating a know amount of water. The first figure of merit, F1 is defined as the ratio of optical efficiency to heat loss factor (preferably >0.12). The second figure of merit, F2 involves the measurement of temperature of known amount of water heated during the course of the day. It should be as large as possible.

Solar dryer: The natural sun drying is simple and economical but suffers from many drawbacks such as: there is no control over the drying rate, the crop may be over-dried resulting in discoloration, loss of germination power, nutritional changes and sometimes complete damage there is no uniform drying in case of slow drying there can be deterioration of food due to fungi and bacteria the rain and dust storm may damage the crop since in open drying there is no protection there will be considerable damage due to birds, insects etc. in the open sun. Solar energy is all the more effective for food drying because of the following reasons: Solar energy is diffuse in nature and provides low grade heat. This characteristic of solar energy is good for drying at low temperature, high flow rates with low temperature rise. The intermittent nature of solar radiation will not affect the drying performance at low temperature. Even the energy stored in the product itself will help in removing excess moisture during the period of no sun. Solar energy is available at the site of use and saves transportation cost. The high capital cost of solar dryers can be compensated if the dryer is used for drying other products also.

Basics of solar dryer: Drying or dehydration of material means removal from the interior of the material to the surface and then to remove from the surface of the drying material. In natural sun drying where the product is directly exposed to the sun in the open air, the necessary heat required for moisture removal is supplied from the sun and a little from the ambient air and the wind and the natural convection disperse water vapor. While in the convection type of solar dryers, a stream of preheated air from solar energy supplemented by auxiliary energy is allowed to pass through the product which supplies the necessary heat for moisture removal from inside to outside and also carries the moisture. The rate of moisture movement from the product inside to the air outside differs from one product to another and very much depends on whether the material is hygroscopic (grains, fruits etc.) or non-hygroscopic (textile materials).Removal of moisture from the surface depends on the difference between the vapor pressure of the moisture in the grain and that of the drying air. Considering the operational modes and practicability of dryers, dryers are classified into three types: the direct type or natural convention type dryers; mixed type dryers and indirect type dryers or forced circulation type dryers.

Natural convection or direct type solar dryers: These dryers appear to be more attractive for use in developing countries since these do not use fan or blower to be operated by electrical energy. Moreover, they are low in cost and easy to operate. the problems with these dryers are: slow drying, not much control on temperature and humidity, small quantities can be dried and some products change color and flavor due to direct exposure to sun. A simple natural convection type solar dryer is cabinet dryer.

Mixed mode type dryer: In the mixed-mode type of dryers, the solar air heater without any fan along with the drying bin is used. The flow of air is generally by natural convection. This dryer named as 'rice dryer' consists of a simple air heater, drying chamber and a tall chimney used to increase the convection effect. The height of the chimney and the hot air inside it creates a pressure difference between its top and bottom thereby creating forced movement of air through the rice bed to the top of the chimney.

Forced circulation type dryers: In these dryers some kind of blower is used for the circulation of air which is either operated electrically or mechanically. Such dryers are more efficient, faster and can be used for drying large quantities of agricultural products. These forced circulation type dryers are also categorized as direct mode forced circulation dryers and indirect mode forced circulation dryers. Direct type dryers are generally used for timber drying. In indirect mode forced circulation dryers temperature can be controlled. Indirect type or forced circulation dryers are very efficient, can be used at low as well as high temperatures and for drying large quantities of agricultural products. These dryers are of bin type, tunnel type, belt type, column type or rotary type.

Solar water heater: Water heating system through the radiation of solar light is another unique system by which it is possible to heat the water up to the extreme limit. Basically the solar collectors are the main source that is responsible to heat the water in a heating system. The main technology behind it is that a solar collector absorbs the solar radiation falling on it, converts this energy in to heat and transfers this heat to water or air flowing through the collector. There are some kinds of collectors those are used for heating water. They are: a) Flat plate collector; b) Evacuated tubular collectors and c) Concentrating collectors. Now we can discuss about various kinds of collectors. First start with flat plate collector; the flat plate collector consists of an absorber plate with channels attached to the plate. Not all of the solar energy falling on the collector is transferred to the water in channels, as some heat is lost to the surroundings. Methods of reducing heat losses include-putting the absorber plate in an insulating box placing extra insulation behind the absorber plate to reduce heat loss from the back of the plate placing a transparent cover the box for two reasons: it prevents the wind from blowing over the hot absorber plate and cooling it and it transmits solar energy but prevents the heat radiated by the absorber plate from escaping (i.e., the greenhouse effect). The construction of flat plate collectors had designed to ensure the object that the flat plate collector consists of an absorber plate with channels attached to the plate. Not all of the solar energy falling on the collector is transferred to the water in channels, as some heat is lost to the surroundings. Methods of reducing heat losses include- putting the absorber plate in an insulating box and placing extra insulation behind the absorber plate to reduce heat loss from the back of the plate. Placing a transparent cover the box for two reasons: it prevents the wind from blowing over the hot absorber plate so that; it does not cool it and transmits solar energy but prevents the heat radiated by the absorber plate from escaping (i.e., the greenhouse effect). We have to keep in mind that there are some requirements for good heat transfer from the plat to the liquid in the riser tubes- the plate material should conduct heat well; a good thermal contact should exist between the plate and the tubes. In general aspect the spacing between riser tubes is normally measures near about around 150 mm ; the tubes are mostly 13mm in diameter, a header tube measures 25 mm, the transparent cover is 30-40 mm above the absorber plate and the insulation behind the absorber plate is between 25 and 50 mm thick. The construction material has their own requirements such as the absorber plates should have a high thermal conductivity- copper and aluminum have been popular choices for absorber plates. While coatings the absorber plate we have to remember that that it absorbs a large fraction of the solar radiation falling on it and it does not deteriorate at the high temperatures which may be reached by the plate. Normally black paint is most common to paint over it. This is because the black paint absorbs no less than 96% of sun radiation. Special surface can be prepared for the purpose. This special surface is called selective surface which is prepared by chemically treating the plate. But one disadvantage is there that it is more costly than the black paint surfaces. Due to this, special surface is used only in some particular case when high temperatures are required from collectors. Insulation is required in the water heating system. The main purpose of insulation is to withstand the temperatures achieved within the collectors as it has a low thermal conductivity. There are boxes to support the absorber plate, insulation and transparent covers. These boxes are known as collector boxes. These boxes have another job which is to protect up the collector components from the unfavorable climate. There is required for a transparent cover. For the reason glass is common material to use. Glass has some uniqueness for getting used as a cover of the heater system. Glass is durable, strong and does not degrade in sunlight. Added advantage of not transmitting the radiation emitted by the hot absorber plate. It is observed that locally made glass has relatively high iron content, leads to significant absorption of solar radiation and a 3 mm sheet of local window glass may only transmit 80-84% of the radiation falling on it. But it has been observed that through low iron glass up to 92% of the solar radiation may be transmitted. Tempered glass can also be used on flat plate collectors. Such collectors are easier to install because the glass is much stronger and may even be walked on. Efficiency of flat plate collector can be measured by the formula as:

$$\text{Efficiency} = \frac{\text{Useful heat out of collector}}{\text{Solar energy falling on collector}}$$

$$\frac{\Delta T}{G} = \frac{\text{liquid or air temperature} - \text{temperature of surroundngs}}{\text{Intensity of solar radiation}}$$

There is another type of collector which is known as evacuated (vacuum) tubular collectors. One method of obtaining temperatures between 100^0C and 200^0C with solar energy is to use evacuated tubular collectors. It has been based on the idea with and evacuated tubular collector is to enclose the absorbing surface in a vacuum. By doing this the only heat which can be lost by convection or conduction is via the tube supports and the ends of the tubes. Because of these very low heat losses the collector can operate quite efficiently at temperature between 100^0C and 150^0C. With small concentration behind the tubes, temperatures up to 200^0C can be achieved. In fact it is very efficient, of the sun light's energy hitting the tube's surface, 93% is absorbed, whereas only 7% is lost through reflection and re-emission. The presence of the vacuum wall prevents any losses by conduction or convection - just like a thermos flask. It is an important job to concentrate the collectors. The main reason for using concentrating collectors is to obtain higher temperatures. It has been observed that it is possible to get a temperature of up to 3800^0C by proper concentration. Now this is the time to know about the hot water storage tanks. Where solar collectors are used to produce hot water (e.g., domestic hot water systems, space heating systems, industrial process heating applications), the simplest method of storing energy is the use of a well insulated tank of water. In comparison to most other common materials, water stores a large amount of energy for each degree increase in temperature. Another favorable property of water is the variation of the density of water as the temperature changes. Hot water is less dense than cold water. Therefore in a tank containing water at different temperatures, the hot water will tend to rise to the top and the cold water will tend to sink to the bottom. The effect is termed stratification. This tank has a unique design to build up. While to design this tank one has to keep remember the existences of following features of the tank: the water for the collectors is taken from the bottom of the tank (hopefully this is the coldest water). The water from the collectors is returned to the top of the tank. The water for the load is taken from the top of the tank. The velocity of water entering and leaving the tank is kept low by the use of ring headers, sparge pipers etc. If an electric in-tank booster is used, the element should be placed in the top half of the tank. In this way the top half of the tank is heated while the bottom half is available for solar heating and lets the conditions become favorable. There is a common rule of thumb for sizing a hot water storage tank for a solar system is to use 75 liters of water for every square meter of solar panel. There are separate usages of hot water for domestic purpose. For the house hold usage purpose the possible solutions can be passive systems and active systems. Thermosyphon water heating system is becoming popular nowadays. Two types of passive thermosyphon solar water heating systems exist- the closed-coupled unit and a system which separated collectors and storage tank. It is possible to heat the water directly and also can possible to use of a heat exchanger. Now we may discuss about the size quantity and timing of a normal heating system. 500 liters of water to be heated daily from 25 to 70⁰C, design the system. The formula to calculate to estimate energy requirement: $mS \Delta T = 500*4180*(40-25)$ kg/day*J/kg-K *K $= 4.91*10^7$ J/day (Monthly average daily radiation data= 4.24 kWh/m²/day); Annual average collector efficiency=40%; Heat on collector/m2 $=4.24*1000*3600$ J/m² $=1.53*10^7$ J/m2; Each 1m2 can produce= $1.53*10^7*40\%$ J/m2/day ;Collector area = Estimated energy requirement/Each 1m2 can produce.

Normally Standard 2m2 collectors are available; No. of collectors= 4

A system needs storage and others.

Typically a system cost is 1.5 times of collector cost.